W9-BPO-242

*"Keep it simple. Short formulas are the right ones. Long formulas are usually wrong."*

# WELDING POWER HANDBOOK

## A Basic Manual on Theory and Use of Arc Welding Power Supplies

A. F. Manz
Project Scientist
UNION CARBIDE CORPORATION
Linde Division
Tarrytown, N. Y. 10591

Permission has been granted by Union Carbide Corporation for the reprinting of this book by the American Welding Society.

# PREFACE

This book provides a source of operating principles and useful data to the user of electric arc welding equipment. Emphasis is placed on principles of operation and not specific details of commercial equipment operation. If you are looking for operating details of specific pieces of welding equipment, this book is not for you.

It is divided into two parts. The first part discusses basic theory. The second part discusses practical applications of the fundamentals discussed. The first part reviews the fundamentals of electricity necessary for understanding the operation of arc welding systems. Rules of thumb are used where possible. A few of the rules of electrical engineering are bent, but not broken. In the second part, emphasis is placed on Tig and Mig (non-consumable and consumable electrode) welding systems. Some elementary knowledge of physics and electricity, as taught in high school or as taught in a vocational school, is helpful.

Chapters II thru V, of Part I, were added to make the Handbook complete. The material in these chapters is more detailed and complex than required for an understanding of Part II. They may be omitted by the less technically inclined reader. Chapters XII and XIII on MIG welding power supplies are a bit more theoretical than the other chapters in this book. A special effort has been made to explain the operation of MIG welding systems.

# CONTENTS

Preface

# PART I

CHAPTER I

# INTRODUCTION TO ARCS
# AND ELECTRICITY

CHAPTER II      # DC FUNDAMENTALS

CHAPTER III      # SINGLE PHASE
# AC FUNDAMENTALS

CHAPTER IV       THREE PHASE
ALTERNATING CURRENT

CHAPTER V      ARC FUNDAMENTALS
AND ELECTRICAL MODELS

PART II
CHAPTER VI     POWER SUPPLY
CONSTRUCTION FUNDAMENTALS

CHAPTER XI      **COVERED ELECTRODE
WELDING POWER SUPPLIES**

CHAPTER XII      **MIG WELDING
POWER SUPPLIES**

CHAPTER XIII      **SHORT CIRCUITING
TRANSFER MIG
WELDING POWER SUPPLIES
(Short Arc Power Supplies)**

CHAPTER XIV    **FLUX-SHIELDED ARC WELDING POWER SUPPLIES**

# APPENDIX

## TABLES

## CHARTS

# PART I

# CHAPTER I

# INTRODUCTION TO ARCS
# AND ELECTRICITY

## What is an Arc?

For all practical purposes, a welding arc can be thought of as something which changes electrical energy into heat. Figure I-1 illustrates the conversion concept. Although the temperature of an arc is high enough to melt any metal; in order to make a weld, the correct amount of heat is also needed. The amount of heat which an arc produces from electricity depends upon a great many things. One of the most important things is the arc current.

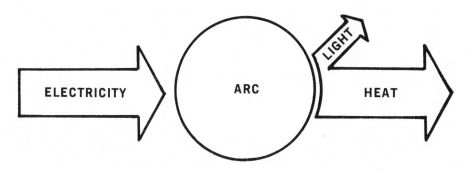

FIGURE I-1: An Arc Is a Conversion Device

When the arc current is increased, the amount of heat the arc produces is increased and vice versa. Another important thing which controls arc heat is arc length. Changes which are made in the length of an arc will cause changes in the amount of heat available from the arc. Consequently, the welding ability of an arc depends upon control of both the arc current and arc length. Most of the time, the arc current is controlled by the welding power supply and the arc length is controlled by the welder.

## Arc Heat

When the *length* of the arc is held *constant,* the amount of heat an arc produces depends upon the current. When a steady current is used, the arc heat is steady. When a pulsating arc current is used, the arc heat pulsates. This concept is shown in Figure I-2.

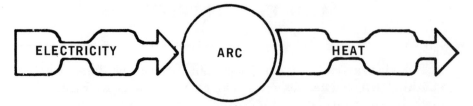

FIGURE I-2: Pulse Power Is Converted to Pulse Heat

The arc heat changes at the same time as the current changes. There is no time lag between the change in heat and the change in current. Neither is there any heat storage. The heat starts and stops with the arc current. These features of an arc make arcs ideal for use where fast heat response is needed, such as in spot welding, as well as where steady heat is needed.

## Parts of an Arc

All arcs have three main electrical parts. These are:
1. Anode - the positive end of the arc
2. Cathode - the negative end of the arc
3. Plasma - the middle of the arc

An arc is usually defined as an electrical discharge between two electrodes (terminals). The anode forms on the positive electrode, the cathode forms on the negative electrode and the plasma is formed in the space between the electrodes. The electrodes (terminals) themselves are not considered as part of the arc.

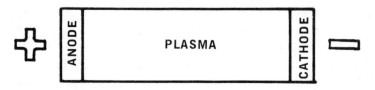

FIGURE I-3: Parts of an Arc

The electrodes merely provide a place for the arc to attach itself. The parts are illustrated in Figure I-3.

Anodes and cathodes are very small and thin. You cannot see them very easily. The plasma is big and bushy. It is the part that is normally visible and looks like a flame. When an arc is formed

2

between two horizontal electrodes, it arches upward. The term *arc* probably came from the term *electrical arch,* which was used by early experimenters to describe their work with arcs between horizontal electrodes. This is shown in Figure I-4.

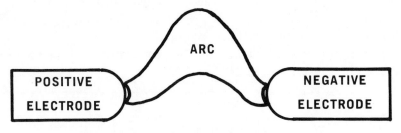

FIGURE I-4: The Horizontal Electrical Arch (arc)

**Current Flow**

Electrical current is the flow of negative charges (electrons) in a wire. When the negative charges (electrons) flow in the same direction all the time, the current is called DIRECT CURRENT or DC. When the electrons change direction back and forth (first flowing one way, then stopping and flowing in the opposite direction), the current is called ALTERNATING CURRENT or AC. The amount of current (electrons) flowing in a circuit is measured in amperes. The *ampere* is discussed in more detail later in this chapter.

When Benjamin Franklin made his famous kite experiment, he decided that current flow is from the POSITIVE to the NEGATIVE terminals in a circuit. At the time it seemed to be a

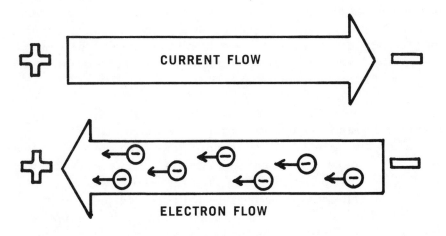

FIGURE I-5: The Conventional Directions of Current and Electron Flow

logical choice. However, he picked the wrong direction. The electrons (negative charges) really flow from the negative to the positive terminals in a circuit. There has been confusion ever since it was discovered that the electron flow is really opposite to the direction of current flow. However, we still use Ben Franklin's direction of current flow. The accepted convention is shown in Figure I-5. Just remember that current flow is *opposite* to electron flow and that current flow is from the positive toward the negative terminal in the circuit.

### Electrons
Although it isn't 100% technically correct, it is convenient to think of electrons as little negative charges that can move about freely in a circuit. They can be thought of as being piled up at the negative end of a circuit, waiting to flow to the other (positive) end of the circuit. The positive terminal does not have enough electrons, the negative terminal has too many. When the two terminals are connected to each other by wires, the negative charges travel to the positive terminal.

Power supplies can be thought of as sources of more electrons. As long as a power supply is connected to a circuit, a negative terminal can never use up its surplus of electrons, and a positive terminal can never receive too many electrons.

### Welding Polarity
The direction of current flow influences the melting efficiency of the welding arc. One arc polarity is hotter than the other. Consequently, the control of the polarity in a welding setup is very important. There are two welding polarities. These are the *straight polarity* and the *reverse polarity* connections. They are discussed below and represented in Figure I-6.

### Straight Polarity (electrode negative—work positive)
When the current in a welding arc flows from the work to the torch (electrode), the arc is called a straight polarity arc. When arc welding was first being developed, straight polarity was used most of the time. This was done because with straight polarity it was possible to melt the electrode faster than with the opposite polarity. It was thought to be the best way or "straight way" to do the job. It is abbreviated as SP.

**Reverse Polarity** (electrode positive—work negative)

When the current flows from the electrode to the work, the arc is called a reverse polarity arc. With this connection, the electrode does not melt as fast as with straight polarity (SP). The reverse polarity connection came into use after the straight polarity connection. Since it is opposite to the straight polarity connection, it is

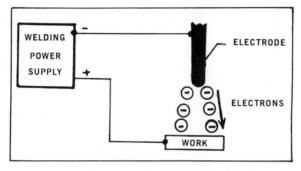

D.C. Straight Polarity Machine Connection

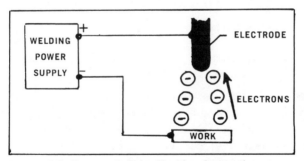

D.C. Reverse Polarity Machine Connection

FIGURE I-6 (a): Definitions of Welding Polarities

called the "reverse" polarity connection. It is abbreviated as RP.

Here is a simple way to remember which connection is what polarity. Everyone knows that the Congress of the United States has both *SEN*ators and *REP*resentatives. Use *SEN* for *S*traight *E*lectrode *N*egative and *REP* for *R*everse *E*lectrode *P*ositive.

5

REP.  SEN.

REVERSE                                      STRAIGHT
ELECTRODE                                   ELECTRODE
POSITIVE                                    NEGATIVE

FIGURE I-6 (b): Definitions of Welding Polarities

### Some Important Concepts

All welding arcs behave according to the same rules. Many of the rules will be discussed in later chapters. Arcs are basically electrical devices and they operate according to the fundamental rules of electricity. The successful use of arcs depends on an understanding of some of these rules. The electrical rules involve resistance, current and voltage. These will be discussed briefly in the following paragraphs.

### What is Electricity?

To most people, electricity is something to be feared. As in most cases, the fear is due to a lack of knowledge. Electricity should be respected, not feared. Every phase of our daily activities is involved in some way with electrical apparatus and equipment. Electricity is essential to arc welding.

As a help in understanding what electricity is about, most of the time you can substitute the word "electron" for the word "electricity." When this is done, the concepts sometimes become a little easier to understand. However, there are some cases where this cannot be done. For example, in some solid state electrical devices the electrical energy is transferred by "Holes." "Holes" are places where electrons are missing. The subject of *electricity* can get complicated. (You can find out more about "holes," and such, by reading one of the many books on *Electronics.*) It is usually easier to use the word "electricity" in a general sense, and not worry about "electrons," "holes," etc. It is usually too much trouble to think about electron flow, anodes, cathodes, etc. We will keep

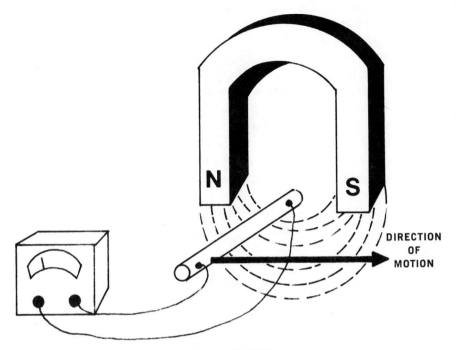

FIGURE I-7: Generation of Electricity

things simple by using terms like *volts* and *amps,* which will be discussed in the following paragraphs.

You cannot see electricity, hear it, taste it or smell it. But you can see its effect, hear its effects, taste its effects and even smell its effects; just as you can see an arc, hear it, feel its heat and even smell the vapors produced near the arc. In the sense that an arc creates heat from electricity, other things can create the effects of electricity with which we normally are acquainted. Sound is obtained from radio and TV speakers, for example. Light is obtained by using a bulb, which also produces some heat. The study of electricity is nothing more than the study of things which convert electrical energy into energy which we can measure and use.

The history of electricity goes back several thousands of years. About 600 B.C., Thales, a Greek, discovered that amber, when rubbed with wool, will attract little splinters of wood and bits of straw. In fact, the word electricity is derived from the Greek word for amber. For 2000 years, that was about the extent of knowledge of electricity. From 1600 A.D. to 1800 A.D., some experimental evidence  of other types of electricity was uncovered. Remember Ben Franklin and his kite experiment? However, it was only in the last 150 years that knowledge of electricity really developed. The discovery that electricity (electrons) can be generated by the *motion* of a *wire* in a *magnetic* field was one of the greatest discoveries of all time. Figure I-7 illustrates the generation of electricity by motion and magnetism.

Unlike most other forms of energy, such as chemical, heat, mechanical, etc., electrical energy (electrons) cannot be stored—it must be generated when it is required; except for the generally trivial case of energy (electrons) stored in a capacitor. It is because electricity *itself* cannot be stored, that we have to have power generating stations, power transmission lines, batteries, etc. Perhaps, some day someone will discover a method by which vast amounts of electrical energy can be stored. But until that time comes, we must depend upon power plants and conversion from energy stored in other forms to provide our electrical needs.

Electricity as we obtain it from the public utility companies is produced by wires rotating in the magnetic field of a generator. Mechanical energy is used to rotate the generators. Since electricity itself cannot be stored, what must be stored then is the energy which produces the rotation of the generator. In the case of some power generating stations, energy is stored in a body of water, which when it is allowed to run downhill, turns turbine type generators. In other systems, fuel is burned in order to create heat, in order to create steam or hot gas to turn a turbine and create mechanical energy for conversion into electrical energy. Atomic power is another example of converting one form of energy into electrical energy. Batteries convert chemical energy into electrical energy and vice versa. However, the energy storage capability of batteries is relatively small. Batteries are usually used when portability is required, such as in automobiles, space vehicles, radios, etc. Photo cells can also turn energy from light into electrical energy. As you can see, there are many ways of producing electricity. A welding arc does not care from where the electricity is obtained, just as long as there is enough of it.

Even motors operate on essentially the same principles as generators. For all practical purposes, motors are made the same way as generators, except with motors electricity goes in and motion comes out, instead of vice versa. In fact, quite often generators can become motors or vice versa. It just depends upon whether you turn the shaft on which the generator wires are wound or put current into the wires and cause the shaft to turn.

Transformers also work according to the same basic principles. They are used in almost all welding equipment. With trans

8

-formers, a *stationary* (primary) coil uses alternating current to produce a *moving magnetic* field. This *moving* magnetic field produces electricity in a second *stationary* (secondary) coil. The relative motion, between the stationary transformer coils and the moving magnetic field produces electricity. Figure I-8 illustrates the transformer concept.

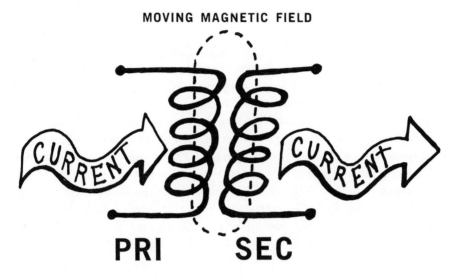

**MOVING MAGNETIC FIELD**

**PRI      SEC**

FIGURE I-8: A Simple Transformer

The ability of wires, moving in a magnetic field, to produce current is described by a property called inductance. Inductance is measured in Henries.* Inductance and its effect on the welding arc is discussed in detail later in the text. The generation of electricity from the simple relationships between motion, magnetism and wires is the basis of electrical power generation. Welding power supply operation is dependent on the concepts of *motion, magnets* and *wires.*

### Definitions

Most dictionaries define a volt as:

"The unit of electromotive force. One volt causes a current of one ampere to flow through a resistance of one ohm."

When you look up the definition of an ohm in a dictionary, it usually says it is "the resistance which will cause one ampere to flow when there is an electromotive force of one volt." If you look

* Named after Joseph Henry, an early American scientist.

up the definition of an ampere, it will be given as "the current which will flow when one volt is applied to a resistance of one ohm." When you put all these definitions together, you haven't found out a thing. It's like a dog chasing its tail.

Definitions are used to define other definitions. In the paragraphs to follow, these terms will be defined with descriptions that have a physical meaning, and that can be associated with a welding arc.

**The Ohm**

The ohm is a unit of resistance. It is a measure of how difficult it is to make electrons flow in a circuit. When the resistance in a circuit is high, more voltage is required to push the electrons through the circuit than when the resistance is low. Every wire and conductor of electricity has resistance. *One* ohm can be defined as the resistance of a piece of wire with certain specific dimensions and measured at a specific temperature. In order to make one ohm of resistance, it is only necessary to duplicate the piece of wire and the temperature. It is not necessary to define the ohm in terms of voltage and current, but just in terms of size and material. Before 1948, the International Ohm was known as the resistance between the ends of a column of mercury 106.3 centimeters long and

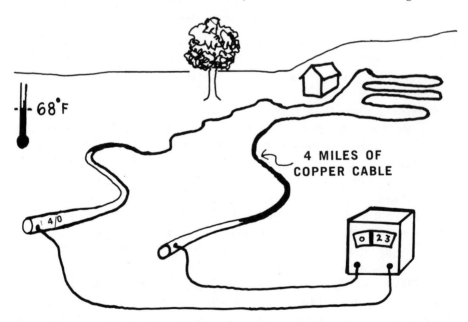

FIGURE I-9: An Example of One Ohm

weighing 14.4521 grams, measured at 0° C. A very precise standard in its time. It takes one volt to push one ampere through this column of mercury. One ohm made with different metal could have different dimensions and temperature. For example, a four mile long piece of 4/0 copper welding cable has a resistance of about one ohm when measured at 68° F (20° C), as is shown in Figure I-9.

Resistivity is used to compare the resistance properties of different materials. It is a measure of a material's resistance to the flow of electrons and is determined by the inherent properties of the material. For example, the resistivity of steel is greater than the resistivity of copper. Table I of the appendix shows the resistivities of a number of materials used for welding. To find the resistance of a specific wire, first determine the wire resistivity from this table, then multiply the resistivity by the wire length and then divide the answer by the cross sectional area of the wire. Remember to use the proper dimensions (units). Do not mix up *feet* with *inches,* or *inches* with *centimeters,* etc.

**The Ampere**
The ampere is a measure of current. Current measures the rate of flow of electrons in a conductor. The more the electrons that flow past a point, the higher the current. A convenient way to describe current is to describe the amount of work it can do. When electrons flow through a resistance, they create heat. The more the electrons (current) that flow, the more the heat that is generated. Resistance welding works by using the resistance heat principle. The heat which is generated is called "I squared R" heat, or $I^2R$ for short. This simply means that the amount of heat can be calculated by multiplying the current by itself and then by the resistance. For example, in a circuit with one ohm of resistance, two amperes would make four watts* of heat (2x2x1). If the current were three amperes and the resistance were two ohms, there would be 18 watts (3x2x2).

It takes a little more than 1000 watts to raise the temperature of one pound of water one degree F in one second. It takes about 570 times that amount of heat (570,000 watts) to melt one pound of steel in one second.

---

* Watt, the electrical unit of power. It measures the rate of work done by electricity. See Page 24.

### The Volt

The definitions of an ohm and an ampere discussed in the previous paragraphs make it easy to define a volt as "the force required to push one ampere through one ohm." Voltage is the pressure that pushes the electrons through the resistance. More pressure (volts) is needed to push the current when the resistance is high. Whenever more current is required in a welding circuit, the voltage must be increased. When the voltage *cannot be* changed because of power supply limitations, the current can be increased by decreasing the resistance of the circuit. The volt, the ampere and the ohm are related by Ohms Law. A thorough discussion of Ohms Law occurs in Chapter II.

# DC FUNDAMENTALS

### Introduction

A great many welding systems use direct current as the source of arc power. The behavior of a direct current arc depends on the behavior of the circuit, in addition to the welding technique of the operator. This chapter deals with the principles behind the control of the current and voltage in the welding system. The operation of welding arcs is discussed in the chapters of Part II.

### The Welding Circuit

In any welding system, the current (electrons) must travel in a complete loop from the power supply to the arc, and back to the power supply. The circuit loop can be completed between the power supply and the arc by cables connected to the torch (one end of the arc) and to the work (other end of the arc). In the example shown in Figure II-1, current is arbitrarily shown as flowing in a clockwise direction. The electrons flow in the opposite direction. (See Figure I-6). This represents the reverse polarity (RP) connection. The current flow is down from the torch to the work. When the current flows in a counter-clockwise direction, or up from the work to the torch, it is known as the straight polarity (SP) connection.

In some systems, such as in field erection work, there is no cable between the work and power supply. Instead, the work is connected to a water pipe or a steel girder, or the work itself completes the circuit.

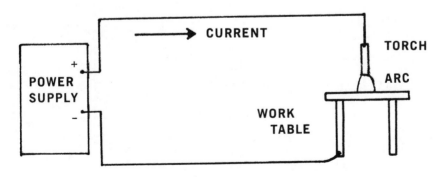

FIGURE II-1: A Typical DCRP System

As is shown in Figure II-2, the circuit loop is completed through a pipe or girder back to the power supply, which is also connected to the same pipe or girder. The current flows through the pipe just as if it were a cable.

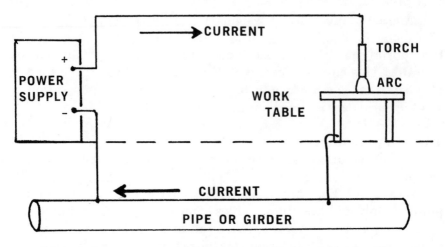

FIGURE II-2: A Welding Circuit Completed with the Help of a Pipe

The simplified circuit diagram shown in Figure II-3 can be used to represent both circuits shown in Figures II-1 and 2.

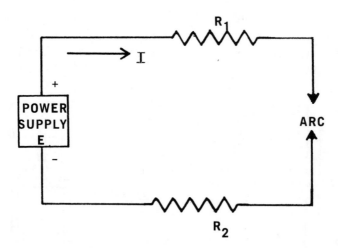

FIGURE II-3: Schematic of Figure II-1 and II-2

14

## Some Symbols

In the circuit of Figure II-3, the resistance of the cables is represented by the ZIG-ZAG symbols ($R_1$ and $R_2$). A resistor is shown in Figure II-4(a). The Zig-Zag symbol is used for a resistor because it makes it seem harder for the electrons to flow through the circuit.

Usually, the power supply is shown as a battery in circuit diagrams, because batteries deliver direct current. The symbol for a battery is shown in Figure II-4(b). One end of the battery has a positive terminal and the other end has a negative terminal. The alternate long and short bars in the battery symbol represent the plates in a battery. The long bar represents the positive end of the battery. The short bar represents the negative end of the battery.

**RESISTOR**                                **BATTERY**

(a)                                          (b)

FIGURE II-4: Typical Electrical Symbols

The welding cables which connect the power supply to the arc may be represented as resistors, since all cables have resistance. The schematic of a welding power supply, an arc and the connecting cables is shown in Figure II-5. The letter (E) represents a power source. In this case it is the battery (power supply) voltage. The (I) represents the welding current, ($R_1$) and ($R_2$) represent the cable resistance and (V) is the arc voltage.* The arrows drawn through the symbols show that the device represented by the symbol is adjustable. Some welding circuits use adjustable resistor boxes.

---

* E, I, R and V are the traditional symbols used to represent electrical devices. E is the abbreviation for *E*lectromotive *f*orce, otherwise known as EMF. I is the abbreviation for current *I*ntensity. R & V are the abbreviations for the *R*esistances and *V*oltage of things which use electrical energy.

15

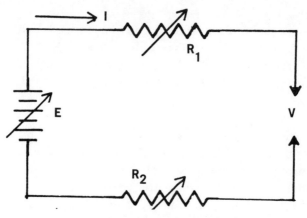

**FIGURE II-5: A Simple DC Circuit**

## More Circuits

In a welding circuit, or for that matter any circuit, the voltage put out by the power supply is used up by the current as it flows (is pushed) around the circuit loop. Think of voltage as a pressure which pushes the electrons around a circuit loop. Some of the voltage (pressure) is used by the arc, the rest is used by the remaining circuit resistance as the current is pushed through the cables and connections. The characteristic of circuit loops to use voltage is described by the following general rule.

> GENERAL RULE II-1: All the voltage *must* be used up in forcing the current to flow around the circuit loop.

The rule is true, for if there were any voltage left over, it would push more current through the circuit until the voltage was used up. Also, if the voltage had been used up *before* the circuit loop was finished, there wouldn't have been enough voltage left to force the current through the remainder of the circuit.

There is another general rule regarding current. It is based on the fact that whatever current leaves the power supply must return to the power supply.

> GENERAL RULE II-2: The current is the same at all places in a single circuit loop. When a circuit branches off, the current flowing out of the branches must be equal to the current flowing into the branch point.

16

This is shown in Figure II-6. It does not make any difference in which direction the current flows or if the current is AC or DC.

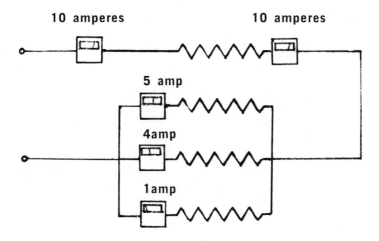

**10 amperes**          **10 amperes**

5 amp

4amp

1amp

**FIGURE II-6: Conservation of Current**

General Rule II-2 for current is true, because when electrons (current) flow into a circuit, they must come out of the other end of the circuit. When there is less current coming out at one end of a circuit than went into the other end, it means that there is a branch circuit somewhere else which was not accounted for. What goes in, *must* come out!

### Ohm's Law and Voltage Drops

Whenever current flows through a resistor, the voltage, which pushes the current, can be measured across the resistor. The voltage is referred to as a "voltage drop" in the same sense that pressure drops in a pipe when water is flowing. Ohm's Law says that the voltage (E) is equal to the current (I) multiplied by the resistance (R), or:

(II-a)          $E = (\text{Current}) \ (\text{Resistance}) = I \times R$

Example II-1:

**Find:** The voltage drop across a stick electrode current adjusting resistor box, when the box is set to 0.2 ohms and the current is 200 amperes

**Solution:** From equation II-a and the data above:
$$E = (200) \times (0.2) = 40V$$

**Answer:** The voltage drop is 40 volts.

17

Ohm's Law may be rearranged in two more different forms for convenience.

(II-b)     $I = \dfrac{E}{R}$

(II-c)     $R = \dfrac{E}{I}$

In the example, if the voltage and resistance had been given, the current could have been found by using the E/R version of Ohm's Law. A simple way to remember all of the forms of Ohm's Law is to use the **OHM'S LAW PYRAMID** shown in Figure II-7.

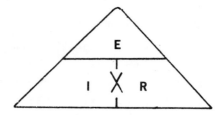

FIGURE II-7: Ohm's Law Pyramid

When you cover any one of the three letters in the Ohm's Law Pyramid with your finger, the other two letters are shown in the correct form of Ohm's Law for finding the value of the covered letter. For example, if the voltage E is covered up, as shown in Figure II-8, then the remaining part of the pyramid shows I X R which represents the correct form of Ohm's Law. Cover up the R and you get E/I etc.

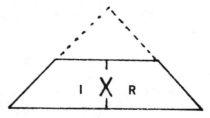

FIGURE II-8: The Voltage Version of Ohm's Law Pyramid

## Ohm's Law and Parallel Resistors

Current control boxes, made by connecting a number of parallel resistors together, are used in stick electrode welding systems to adjust the welding current. The operation of the resistors boxes and their parallel connected resistors depend on the following general rule.

GENERAL RULE II-3: The voltage drop across parallel connected resistors is the same for each resistor.

This rule is true because the parallel resistors are all connected to the *same* terminals. Although the voltage is the same across each of the resistors, the current flowing through the resistors can be different if the value of the resistors is different.

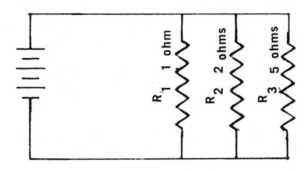

FIGURE II-9: Example of Parallel Connected Resistors

In Figure II-9, three parallel resistors are connected to the same terminals. The current in each of the resistors can be found by using Ohm's Law on each resistor.

Example II-2:

Find: the current in each of the resistors, when the resistors shown in Figure II-9 are connected to a battery delivering 10 volts

Solution: Use the proper form of Ohm's Law as determined from the Ohm's Law Pyramid shown in Figure II-7.

19

$$I_1 = \frac{E}{R_1} = \frac{10}{1} = 10 \text{ amperes}$$

$$I_2 = \frac{E}{R_2} = \frac{10}{2} = 5 \text{ amperes}$$

$$I_3 = \frac{E}{R_3} = \frac{10}{5} = 2 \text{ amperes}$$

**Answer:** The currents are shown in Fig. II-10.

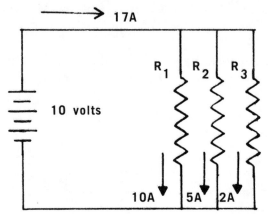

FIGURE II-10: Current Flow of Figure II-9 With a 10 Volt Battery

The current delivered by the battery in Figure II-10 is the sum of all of the branch currents (General Rule II-2). As far as the battery is concerned, it does not care how many resistors are connected in parallel. All it cares about is the 17 amperes total. The 17 amperes could just as well have come from a single resistor which gave the same circuit current.

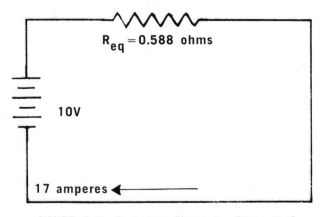

FIGURE II-11: Equivalent Circuit for Figure II-10

The value of the equivalent resistor in Figure II-11 may be found from Ohm's Law as shown below:

$$R_{eq} = \frac{E}{I} = \frac{10}{17} = 0.588 \text{ ohms}$$

By mathematical analysis, it is possible to show that the equivalent resistance $(R_{eq})$ can be found directly from the values of the parallel resistors $(R_1)$ $(R_2)$ and $(R_3)$.

The general rule describing this method is shown below:

> GENERAL RULE II-4: The reciprocal of the equivalent resistance of parallel connected resistors is equal to the sum of the reciprocals of the individual resistors.

When shown in equation form this means:

(II-d)
$$\frac{1}{R_{eq}} = \frac{1}{R_1} + \frac{1}{R_2} + \frac{1}{R_3} \quad \text{etc., etc.}$$

The meaning of General Rule II-4 and the equation above becomes clear when we examine the example of Figure II-10 with 3 parallel resistors of 1, 2 and 5 ohms. From the equation for General Rule II-4.

$$\frac{1}{R_{eq}} = \frac{1}{1} + \frac{1}{2} + \frac{1}{5}$$

$$= 1 + 0.5 + 0.2 = 1.7 \text{ mho*}$$

The equivalent resistance then is the reciprocal of 1.7 or:

$$R_{eq} = \frac{1}{1.7} = 0.588 \text{ ohms}$$

This is the same answer as was calculated in the previous example.

Now that we have discussed some of the theory, let us put the theories to use in a welding example.

* The term *mho* is *ohm* spelled backwards. The mho is a measure of conductance, the opposite of resistance.

Example II-3:

**Find:** the arc voltage when 200 amperes is flowing in the circuit shown in Figure II-12. The welding power supply
has 80V output. The cables connecting the power supply to the arc have a total resistance of 0.05 ohms and a
resistance grid made up of 5 parallel one ohm resistors
is used to determine the final current.

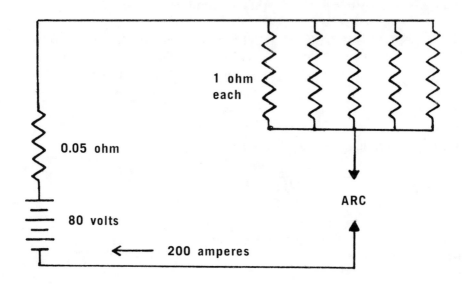

1 ohm
each

0.05 ohm

ARC

80 volts

← 200 amperes

FIGURE II-12

**Solution:** First determine the equivalent resistance of the 5
parallel one ohm resistors using General Rule IV.

$$\frac{1}{R_{eq}} = \frac{1}{1} + \frac{1}{1} + \frac{1}{1} + \frac{1}{1} + \frac{1}{1} = 5 \text{ ohm}$$

$$R_{eq} = \frac{1}{5} = 0.2 \text{ ohms}$$

Now redraw the circuit as shown in Figure II-13, using $R_{eq}$ instead of the 5 parallel resistors.

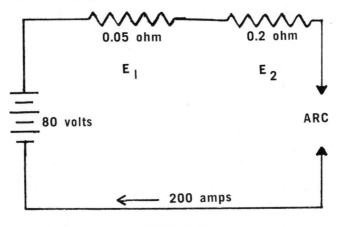

**0.05 ohm**     **0.2 ohm**

$E_1$     $E_2$

80 volts     ARC

← 200 amps

**FIGURE II-13**

Calculate the voltage drops in each resistor by using Ohm's Law.

$$E_1 = (200)\ (0.05) = 10V$$

$$E_2 = (200)\ (0.2) = 40V$$

**Answer:** The total voltage drop in the resistors is 50V, leaving 30V for the arc.

**Ohm's Law and Series Resistors**

The circuit diagrams of some welding systems usually show a line of resistors connected one to another, just as is shown in Figure II-13 of the previous example. A review of the example shows that the following general rule applies.

> GENERAL RULE II-5: The equivalent resistance of series connected resistors is the sum of the resistances.

When shown in an equation, this means:

(II-e)     $$R_{eq} = R_1 + R_2 + R_3 + \text{ etc., etc.}$$

23

The previous example could have used General Rule II-5 to find the arc voltage. The circuit could have been redrawn one more time with the 0.05 ohm resistor and the 0.2 ohm resistor changed to a single equivalent resistor of 0.25 ohms. This is shown in Figure II-14:

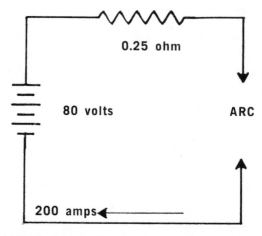

FIGURE II-14: Equivalent Circuit of Figure II-13

The voltage drop across the 0.25 ohm resistor of Figure II-14 is 50V when the current is 200 amperes. This leaves 30V for the arc. The answer is the same as was shown in the previous example.

### Power and Ohm's Law

The amount of heat produced by an arc depends upon the quantity of electrical power which is converted to heat by the arc. Electrical power in circuits is measured in watts. The equation for watts (W) is:

$$\text{(II-f)} \qquad W = EI = (\text{volts}) \, (\text{amperes})$$

For example, a 25V arc operating at 200 amperes converts 5000 watts of electrical power. Resistors also use watts. The current flowing in a resistor causes it to get hot. Since the voltage across a resistor is equal to the current (I) times the resistance (R), it is possible to substitute IR for E in the watts equation above. This gives a second way of writing an equation for power as follows:

$$\text{(II-g)} \qquad W = IR \times I = I^2R$$

From the Ohm's Law pyramid*, it is obvious that (E/R) can also be substituted for I in the watts equation above. This gives a third equation for power.

(II-h) $$W = E \times \frac{E}{R} = \frac{E^2}{R}$$

This equation can also be written as $V^2/R$. The letter $V$ is usually used to represent the *voltage drops* in a circuit. The letter $E$ is used for *voltage sources*. All three of the power equations are based on Ohm's Law. The amount of power generated is frequently referred to by KW, the abbreviation for Kilowatts, kilo means 1000. One KW means 1000 watts.

### Efficiency

Generally, the efficiency of a welding circuit is calculated by comparing the arc power with the power wasted as heat in circuit resistances. Theoretically, if there are no resistance losses in a power supply circuit, it is considered as 100% efficient. Circuits which use resistors for current adjustment are not very efficient.

### Speed of Response

When a welding circuit is turned on, current begins to flow. However, it takes a little time for the current to "build up." The rate at which the current builds up depends upon the circuit's *speed of response*. Some circuits are "faster" than others. Some are slower.

One of the things which control the rate of current build up or speed of response in a circuit is inductance. When there is a lot of inductance in a circuit, the circuit has a slow speed of response and vice versa. The more the inductance, the longer it takes for the current to build up. *All* circuits have inductance. The amount of inductance in a circuit depends on the amount of magnetism which can be produced by the current flowing in the circuit.

### Magnetism and Inductance

Every current carrying wire has a magnetic field set up around it. From the very smallest wire with low current, to large wire with high current. When the magnetic field is large, inductance is large. Anything which increases the magnetic field created by a current increases the inductance and slows the speed of response of the circuit. The larger the magnetic field, the longer

* See Table II of the Appendix for all the versions of Ohm's Law.

it takes the current to build up to its final value. In a sense, the current is held *down* while the magnetic field is building *up*.

The effect of the inductance and its magnetic field can be increased by using more turns of wire and an iron core as shown in Figure II-15. These things make a stronger magnetic field and consequently make more inductance.

FIGURE II-15: Inductance Symbol

Figure II-16 shows an inductor and its effects on the circuit speed of response. With no inductance in the circuit, current "jumps" to its full value as shown in Figure II-16(a). With inductance added, the current "builds up" slowly as shown in Figure II-16(b).

Note that the current takes longer to get to the 100% level when inductance is present. As you might suspect, the same thing happens, but in reverse, when a circuit is turned off. The inductance keeps the current flowing while its magnetic field is disappearing.

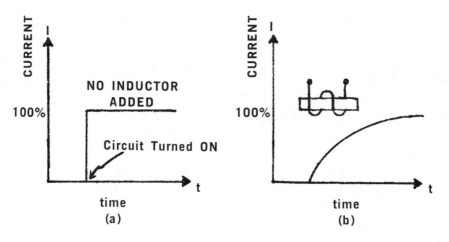

FIGURE II-16: The Effect of Inductance on Current Rise

26

The no inductance case is shown in Figure II-17(a). Note the sharp current decay. Figure II-17(b) shows the slow rate of fall when inductance is present.

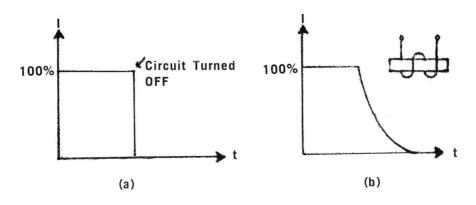

(a)                                                        (b)

FIGURE II-17: The Effect of Inductance on Current Fall

An examination of the current build up and decay shown in Figures II-16 and 17 and the help of a little intuition, will derive the following general rule:

GENERAL RULE: II-6 *Changes* in current *are opposed* in circuits having inductance.

In the current "turn on" case of Figure II-16, the current was held back, or down. In the current "turn off" case of Figure II-17, the current was kept flowing a little longer. In each of the sketches, the change of current from one level to another was "opposed." The direction of the change did not make any difference. The energy of *opposition* is stored in the magnetic field. The bigger the field, the bigger and longer the opposition. The effect of inductance on current change is important to MIG welding.

### Statics and Dynamics

Previous paragraphs of this chapter speak of different types of circuit behavior. All circuit behavior can be divided into two general categories. These are: 1) Static and 2) Dynamic.

The static behavior category refers to things which have nothing to do with speed of response or time. The static behavior category relates to things such as resistors, batteries and things where time is not important. The dynamic behavior category refers to the speed of response of a circuit to changes. It depends on things like inductance, magnetism and things where time is important. Both the static and dynamic behavior of circuits have an effect on the welding system. These effects are discussed in the chapters of Part II.

CHAPTER III

# SINGLE PHASE

# AC FUNDAMENTALS

## Introduction

Except for the power made by engine driven welding generators and batteries, all welding power starts out as alternating current. In fact, at least 90 percent of the electrical energy used throughout the world is generated as AC. The two main reasons for this are that AC can be "transformed" to higher or lower voltages and it can be economically transmitted over long distances. When AC is required for welding, the high voltage power which is delivered by a utility company is converted to the proper welding voltage by transformers.

Alternating current is not as simple to understand as Direct current. Things are constantly changing. This chapter discusses the principles behind the control of alternating current and voltage in a welding power supply.

## What is Alternating Current?

As its name implies, and as was discussed in Chapter I, AC is current which changes its direction. First, it flows one way in a wire, then it flows in the opposite direction. Back and forth, alternately. In the United States, it does this 60 times in one second. People are used to calling the frequency of alternation either 60 (or 50) cycles per second, or 60 CPS.* Actually, the correct term is 60 Hertz or 60 Hz. The name Hertz is the official international designation of frequency. Most U. S. manufacturers now use Hz, instead of CPS, on name plates and in their literature.

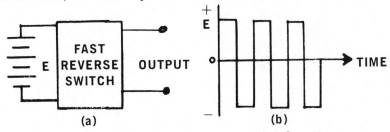

(a)  (b)

FIGURE III-1: Making Square Wave Alternating Current

* 50 cycle power is used in many areas of the world.

## Wave Shape

Alternating current is meant to have a *sine wave* shape, unless otherwise specified. For example, square wave AC could be obtained from the circuit shown in Figure III-1(a).

Each time the fast reverse switch is operated, the output voltage changes direction. First, the voltage is "full on" in one direction, then it is "full on" in the opposite direction. The wave shape shown in Figure III-1(b) has "square" corners.

Most generators create a gradually changing voltage called a sine wave. Instead of being "full on" and suddenly reversing, as shown in Figure III-1, the voltage builds up slowly to a positive peak and then gradually reduces to zero. At zero, the polarity reverses and the voltage slowly builds to a negative peak. Then the voltage reduces to zero and the process repeats itself. This is illustrated in Figure III-2.

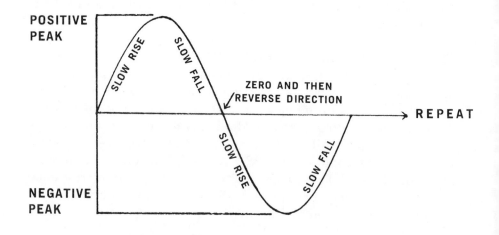

FIGURE III-2: One Cycle of a Sine Wave

Each complete build up and reversal is called one cycle of AC. The shape of the rise and fall curve is special. It is called a sine wave. The sine wave shows up in other places, not only in electricity. In mechanical systems for example, there can be sinusoidal motions. Commercial AC, in the United States and Canada, is a 60 Hertz sine wave.

## Phase Angle

The *phase angle* is used to locate points on a sine wave. When an AC welding arc is turned off, the current probably will not be at *zero* on the sine wave. Chances are that it will be somewhere on the rise or fall portion of the wave. In fact, whenever an AC circuit of any type is turned "on" or "off" the sine wave delivered by the utility company will probably be at someplace other than zero. Special electrical devices are required for "zero" switching.

In addition to wanting to know the switching point on a sine wave, there are other reasons for locating different points on the sine wave. Points on a sine wave may be located by specifying the number of "degrees" they are from the zero point (where the wave starts to rise). The number of degrees is called the *phase angle*. Each complete cycle of AC is divided into 360 equal parts, called *degrees*. Half way through a cycle is the same as 180 degrees (180°). Sometimes this is referred to as the 180° phase angle. Figure III-3 shows some of the important points on a sine wave and the phase angles which locate the points. A quarter of a cycle has a 90° phase angle. Three quarters of a cycle has a 270° phase angle, etc.

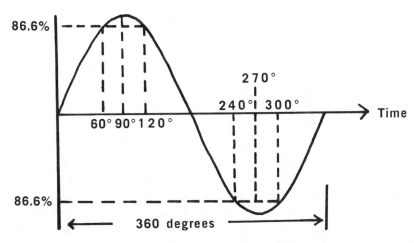

FIGURE III-3: Some Important Phase Angles and Reference Points on a Sine Wave

## Phase Shift

Phase shift is very important in the operation of AC welding circuits because it determines *when* things happen. Sine wave currents and voltages are manipulated inside a welding power supply. The amplitudes of the sine waves may be changed and some waves may be shifted back and forth with respect to one another. For example, one sine wave may be delayed from starting its rise and fall, while another wave may start early. The sine wave that is delayed is said to LAG. The wave that starts early is said to LEAD. Circuits are referred to as being "Lagging" or "Leading."

The amount by which one sine wave *leads* or *lags* another is called "phase shift." Phase shift is also measured in degrees. For example, when one circuit begins 60 degrees later than another, it is said to have a 60° phase shift-lagging.

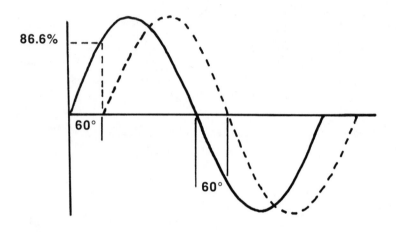

FIGURE III-4: Example of a 60° Lag

The dashed curve in Figure III-4 is shown 60° behind the solid wave. The phase shift is 60° LAG. To the eye it looks as if the dash curve is ahead of the solid curve, but don't let your eyes fool you. The solid curve was already at 86.6% of its peak value when the dashed curve began its rise above zero. Therefore, it is said to be *ahead* of the dashed curve; in other words, the dash curve *lags*. The dashed curve in Figure III-5 started out 60° earlier than the solid curve. It was already at 86.6% of its peak when the solid curve began its positive rise. The dashed curve is said to have a 60° phase shift-leading. The amount of phase shift is usually compared (referred) to the sine wave describing the terminal voltage of the AC supply.

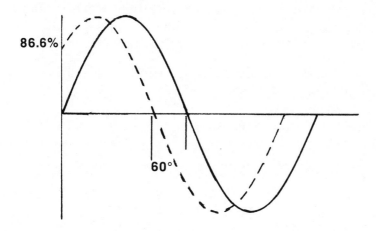

FIGURE III-5: Example   of a 60° Lead

## The AC Ampere

It is easy to measure direct current because it doesn't change, but how do you measure alternating current? It is always changing. An AC sine wave goes up and down 60 times a second. How can it be measured? One way is to measure the amount of work the current can do.

In a DC circuit, one ampere will create one watt of heat (work) in a one ohm resistor. By definition, when one watt of heat (work) is created in an AC circuit, by current flowing in one ohm of resistance, one ampere is said to be flowing. The AC ampere which creates one watt in one ohm has a special name. It is called the

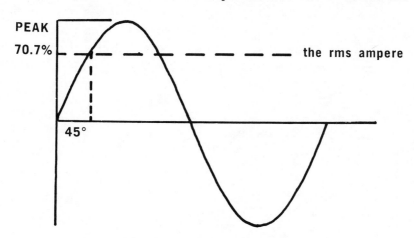

FIGURE III-6: Definition of the AC Ampere

33

*rms* ampere. The term rms is an abbreviation which stands for root-mean-square*. Normally, when someone says an AC circuit has 10 amperes flowing, he automatically means 10 amperes rms.

The rms ampere turns out to have the same heating effect as a direct current 70.7% of the peak of the AC sine wave. This is illustrated in Figure III-6.

The rms value of any sine wave is its *effective* value. It is the same value as a point on a sine wave 45° after it has started.

GENERAL RULE III-1: The AC (rms) ampere, flowing in a resistor will produce heat at the same rate as a DC ampere.

### The AC Volt

An AC volt is defined as the voltage produced by one AC ampere when it flows through one ohm of pure resistance (no inductance, capacitance or stray effects). It has the same sine wave shape as the current and it obeys the same measurement rules.

The meter on a welding power supply with a sine wave with a 100 volt peak will read 70.7 volts AC (100 volts times 70.7% ). A power supply with a meter reading 100 volts AC has a 141.4 volt peak sine wave (100 divided by 70.7% ).

### The Addition of AC Currents or Voltages

In direct current circuits with several parallel branches, the currents are added to get the total current.** This is not true in AC circuits except for the special case when the currents are said to be "in phase". Currents are "in phase" when there is no phase shift between them. Two or more *in phase currents* or *voltages* can be *added arithmetically.* Welding power supplies are sometimes connected in *series* or *parallel* to increase their output to a single arc.

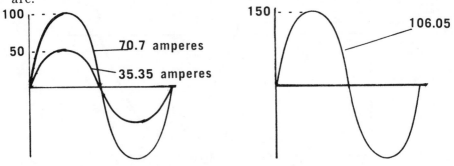

FIGURE III-7: The Addition of "in phase" Currents

* Root-mean-square is a special mathematical term used by Electrical engineers.

** See General Rule II-2, page 16.

In Figure III-7(a), the two currents (100 amperes peak and 50 amperes peak) are in phase (no shift). Ammeters will read 70.7 and 35.35 amperes, respectively. The currents add to give a single current with 150 amperes peak as shown in Figure III-7(b). An ammeter will read a total of 106.05 amperes (the rms value of a sine wave with a 150 ampere peak).

When AC *currents or voltages* are *"out of phase"*, *they do not add arithmetically.* They must be added instantaneously, point by point. It is even possible for two equal sine waves, but out of phase by 180°, to add up to zero.

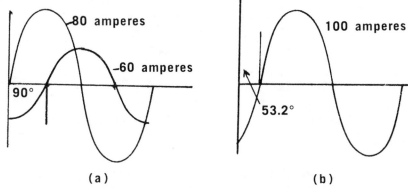

(a)                                    (b)
FIGURE III-8: The Addition of "out of phase" Currents

In Figure III-8, an 80 ampere wave and a 60 ampere wave add up to 100 amperes. It is almost like saying $80 + 60 = 100$. Do not expect AC amperes to add up like DC amperes. Fig. III-8 shows a 90° phase shift between the waves. Note that at 53.2° the 80 ampere wave is the same size as the 60 ampere wave. Since one wave is plus and the other is minus, they add up to zero at 53.2°. This is shown in Figure III-8 (b).

The following General Rule summarizes the requirements for the addition of alternating currents.

GENERAL RULE III-2: The addition of AC currents, or voltages, depends on their phase relationship as well as their amounts.

This general rule is especially meaningful in multiple wire, AC submerged arc welding.

## IN PHASE Circuits, or Circuits with Resistance (R)

Resistive circuits have currents which are always in phase with the voltage. There is no phase shift between current and voltage. When the voltage is at a peak, the current is also at a peak. When current is zero, voltage is also zero. This is shown in Figure III-9.

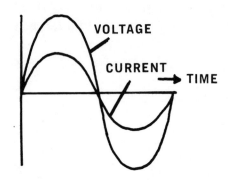

FIGURE III-9: Voltage and Current in a Resistive Circuit

There is no lag or lead in resistive circuits. The current responds instantaneously to changes in the applied voltage.

> GENERAL RULE III-3: Current and voltage are "in phase" in resistive circuits.

## LAGGING Circuits, or Circuits with Inductance (L)*

In Chapter II, General Rule II-6 said, "Changes in current are opposed in circuits with inductance." It is the opposition to current changes that causes the LAG, between current and voltage, in a circuit with inductance. Voltage applied to an inductor tries to make current flow. However, unlike resistive circuit, where the current can flow immediately, the current in an inductive circuit doesn't flow immediately. The current (I) lags the applied voltage (E). When an inductor is perfect (no resistance, capacitance or stray effects), the current lags by *exactly* 90 degrees. As shown in Figure III-10, the phase shift is 90 degrees.

---

* The letter L is the traditional symbol for inductance. It implies the magnetic flux "linkages" responsible for the inductance phenomena.

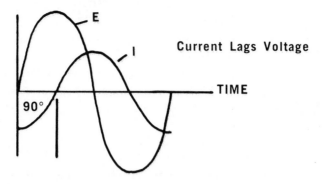

**FIGURE III-10: Voltage and Current in an Inductive Circuit**

The current lag of pure inductance is such that the current just begins to flow when the voltage is at its peak. The current is at its peak when the voltage is zero. As shown in Figure III-10, the voltage can even be reversed and building toward its negative peak while the current is still positive and falling toward zero. The current is zero when the voltage is at the negative peak.

Instead of always saying the current *lags* the voltage, it is also possible to reverse the words and say that the voltage (E) *leads* the current (I) in an inductive (L) circuit. Both are true. However, it is an established convention that an inductive circuit is called a LAGGING circuit.

Here is a simple way to remember *which* leads *what* (whether or not E leads I or I lags E). Think of the name *Eli,* spelled with the capital ELI. In the name ELI, the E is written first. It *leads* L which is written second, which leads the I. In other words, voltage leads the current just as E *leads* I in Eli. In reverse then, I lags E in Eli. Even though both ways are correct, the accepted convention is as stated in General Rule III-4.

GENERAL RULE III-4: Current *lags* voltage by 90° in inductive AC circuits.

### Inductive and Resistive Circuits

Welding power supplies are made from "real" components, not "ideal" or "theoretical" parts. Resistors are not perfect. They have some inductance as well as resistance. Even the smallest piece of wire has some inductance. Conversely, inductors are not perfect either. They have some resistance. All current carrying conductors have some resistance. Real parts have both resistance and inductance.

Therefore, in a "real", not "ideal", welding power supply the phase shifts are not zero degrees (resistive) or exactly 90° lagging (inductive). They are somewhere in between, depending on how much resistance and inductance actually exists. Phase shifts are very important when it comes to AC arcs, particularly when welding aluminum. The effect of phase shifts in AC welding will be discussed in more detail in the chapter on TIG welding

### Current in Lagging Circuits

The amount of current which flows in an inductive welding circuit depends on the "impedance" of the circuit. "Impedance" is used in the sense that inductors *impede* current flow, just as resistors *resist* current flow. The impedance of an inductor is usually represented by the symbol $X_L$.* Just as resistance (R) is measured in ohms, the impedance of an inductor ($X_L$) is measured in ohms. As you might expect, the impedance $X_L$ can be substituted in place of R in the equations for Ohm's Law.

Example III-1:

**Find**: The current in an AC circuit with a source of 20 volts (rms) and an inductive impedance of 0.1 ohms. See Figure III-11.

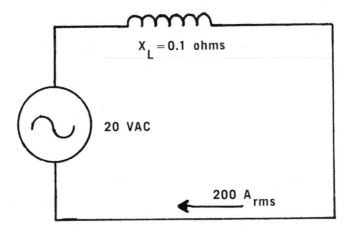

$$X_L = 0.1 \text{ ohms}$$

20 VAC

$$200 \text{ A}_{rms}$$

FIGURE III-11: Circuit of Example III-1

---

* The impedance of condensors is represented by $X_c$, that is why the subscript L must be used here with the letter X.

**Solution**: From the Ohm's Law Pyramid, we find that $I = (E/R)$ Substitute $X_L$ for R to get $I = (E/X_L)$. $I = (20/0.1) = 200$ amperes (rms)

**Answer**: 200 amperes AC

A WORD OF CAUTION IS NECESSARY AT THIS POINT. The current flowing in a resistor *will* produce heat. The electrical energy which is converted to heat in a resistor is lost, it cannot be used to run a motor or an arc. The current flowing in a perfect inductor *will not* produce heat. All of the electrical energy is converted to a magnetic field surrounding the inductor. The energy is recovered when the field returns to zero. *Do not use $X_L$ in Ohm's Law to calculate AC watts.* Only things with resistance have watts.

Instead of watts, an inductive circuit is said to produce "voltamperes." In Example III-1, the inductance is said to have (20 volts) x (200 amperes) or 4000 voltamperes, not 4000 watts. Instead of writing voltamperes out in full, the abbreviation VA is used most of the time. Just as in the DC watts case, a K may be used as an abbreviation for kilo. The 4000 voltamperes is generally written as 4KVA. Many manufacturers of welding power supplies specify the KVA rating in their literature and on nameplates.

### The Effect of Frequency on Inductance

The impedance of an inductor depends on the frequency of the AC. Inductors have zero impedance when the frequency is zero (Direct Current). They do not effect the amount of steady state direct current. The impedance of an inductor rises as the frequency goes up. The increase in impedance is a consequence of General Rule II-6. The greater the rate of change of current, the greater the opposition. When current is *constant* and *does not alternate,* the inductance does not have "changes in current (that) are opposed."

As the frequency of alternation is increased, the rate of change of current is increased. Consequently, the inductance opposes the change with greater strength. The relationship between circuit frequency (f), and the impedance of the inductor ($X_L$) is shown in the following equation.

III-a
$$X_L = 2\pi fL$$

where: $X_L = $ ohms

$2\pi =$ a constant, which is required because of the sine wave shape of current

$f =$ frequency in Hertz, or cycles per second

$L =$ inductance, in Henries*

For 60 Hertz circuits, as in the United States, equation III-a can be simplified to:

III-b $\qquad\qquad X_L = 377L$

Example III-2:

FIND: The current flow through a 0.001 henry inductance when it is connected to a 60 Hz supply with 75.4 volts output. See Figure III-12.

Solution: From equation III-b
$$X_L = (377)\ (0.001) = 0.377 \text{ ohms}$$

$$I = \frac{75.4 \text{ volts}}{0.377 \text{ ohms}} = 200 \text{ amps}$$

Answer: 200 amps

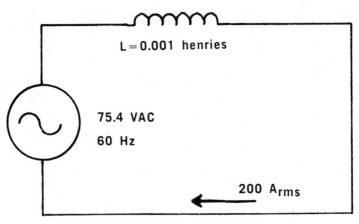

L = 0.001 henries

75.4 VAC
60 Hz

200 A$_{rms}$

FIGURE III-12: Circuit of Example III-2

* Inductance is measured in units called Henries. Henries do not change with frequency.

**Leading Circuits,** or Circuits with Capacitance (C)**

Except in a few cases, capacitors (condensors) are not used to control current in welding circuits. When they are used, it is usually for a secondary purpose such as filtering or power factor correction. In a sense, capacitors are the electrical opposite of in- ductors. For example, inductors oppose changes in current, capacitors welcome the changes. Capacitors block the passage of DC, inductors offer no impedance at all. Inductors block the passage of high frequency current, capacitors offer very little im- pedance (depending on the frequency). The current in inductive circuits lags the voltage, in capacitive circuits the current leads the voltage. In pure capacitive circuits, the lead angle is 90°. The op- posite of General Rule III-4 gives General Rule III-5 for the phase shift in capacitive circuits.

GENERAL RULE III-5: Current *leads* voltage by 90° in capacitive AC circuits.

Here is a simple way to remember the capacitive phase shift rule. Think of the word ice, in capital letters ICE. The I is written first, it leads the C which is written second, which leads the E. In other words, I leads E in ICE. In reverse, then, E lags I in ICE.

With the name *Eli* for inductance, and the word *Ice* for capacitance there should be no problem in remembering the General Rules III-4 and III-5. (The L in Eli and the C in Ice will help you remember which rule is which). As a memory aid, think of *Eli* the *Ice* man.

The discussion of capacitance will be stopped here, inasmuch as capacitance plays only a minor role in most welding circuits. Other details of capacitance will be discussed when necessary in the remainder of the text.

**Power Factor**

Very often, manufacturers of welding power supplies will specify the power factor (PF) of their equipment. The utility com- panies are particularly interested in power factor. Why? The an- swer has to do with power consumption. You will recall that resistance circuits use WATTS and inductive circuits use VOLTAMPERES.* Watts generate heat in resistors, or arcs. Watts are also used to measure the electrical energy that is used in producing motion (motors) and other forms of energy that are con- *sumed.* On the other hand, voltamperes are *not consumed.* They do

** The accepted symbol for capacitance is C.

* Capacitive circuits also use VOLTAMPERES.

not cause energy loss. They do not do work. They are merely stored in the circuit.

Since all circuits are a combination of resistance, inductance and/or capacitance, some measure of their effective use of power is needed. *Power factor* is used to do the job. Theoretically, a circuit with pure resistance is said to have a PF=1. All the electrical power is used up in doing work. A theoretical inductive or capacitive circuit is said to have a PF=0. None of the power is consumed. Actual circuits have a PF between 0 and 1. Inductive and resistive circuits are said to have LAGGING PF. Capacitive and resistive circuits have LEADING PF. Circuits with all three elements (R, L and C) can have either a lagging or leading PF, depending on which is dominant; the inductance or the capacitance.

The PF does not say anything about *how much* power is being used. It is only used to classify the circuit as R, L, or C, or some combination. A power source with a lagging PF=0.96 has very little inductance compared to the resistance. Such a power factor is typical of constant potential MIG power supplies. A power source with a lagging PF=0.6 has quite a bit of inductance. Such PF's are typical of constant current TIG (also covered electrode) power supplies.

The Equation III-c shows how power factor can be calculated.

III-c    $$PF = WATTS / \sqrt{WATTS^2 + VOLTAMPERES^2}$$

Example III-3:

**Find:** the PF of a circuit with 2KW and 2KVA
**Solution:** From Equation III-c and the data above

$$PF = 2KW / \sqrt{(2KW)^2 + (2KVA)^2}$$
$$= 2 / \sqrt{4 + 4} = 2/\sqrt{8}$$
$$= 2 / \sqrt{4 \times 2} = 2/2\sqrt{2}$$
$$= 1/\sqrt{2}$$

**Answer:** 0.707 (lead or lag depending on circuit)

The answer would be the same even if the circuit had 1 watt and 1 voltampere. The PF can also be calculated directly from the resistance (R) and impedance ($X_L$ or $X_C$). This is shown in Equation III-d.

III-d    $PF = R/\sqrt{R^2+X^2}$

**Example III-4:**

**Find:** the PF of a circuit with R = 3 ohms and X = 4 ohms

**Solution:** From Equation III-d and the data above
$$PF = 3/\sqrt{3^2+4^2} = 3/\sqrt{9+16}$$
$$= 3/\sqrt{25} = 3/5$$

**Answer:** PF = 0.6 (lead or lag depending on circuit)

Note in Equations III-c and III-d, that it is the watts component (resistance) that is always divided by the square root to find the PF.

Sometimes, power factor is referred to in terms of an angle. How can that be? As shown in Figure III-13, it is possible to represent R and X by the sides of a triangle.

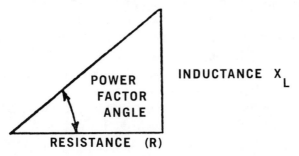

FIGURE III-13: Definition of the Power Factor Angle

The length of the sides are made proportional to the ohms of R and X. The long side (hypotenuse) of the triangle is equal to the $\sqrt{R^2+X^2}$ as is shown in Figure III-14.

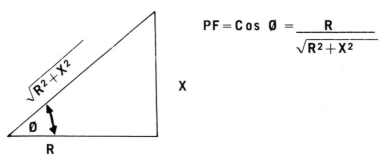

$$PF = Cos\ \emptyset = \frac{R}{\sqrt{R^2+X^2}}$$

FIGURE III-14: Definition of Power Factor

43

In trigonometry, there is another name for $R/\sqrt{R^2 + X^2}$ as defined in Equation III-d. It is called the *cosine* of the angle ($\emptyset$) between R and the hypotenuse. At times, you may see PF = cosine ($\emptyset$), instead of PF equals some number. In Example III-3, the answer could have been written as PF = cosine 45° instead of PF = 0.707.

### Power Factor Correction

There are times when it is desirable to have the PF closer to one. This may be required by the company from which you get your power. A low PF means the power delivered is not used as effectively as it could be. Although inductance and capacitance do not use power (watts), they do draw voltamperes. The power line from the utility company can only deliver a certain amount of current. More cables must be added to carry more current. Therefore, low PF is to be avoided. A high PF makes the most use of the available power line capacity.

PF correction is simple. Remember, it was pointed out that capacitance and inductance are practically electrical opposites? Because they are the opposite of inductors, capacitors can be added to circuits to help reduce the PF angle. This is illustrated in Figure III-15. Capacitors are used to cancel the effects of inductance. An inductor-resistor circuit has a positive PF angle. A capacitor-resistor circuit has a negative PF angle. When both types of circuits and PF are combined, if the angles are equal, they cancel one another. The corrected PF equals one.

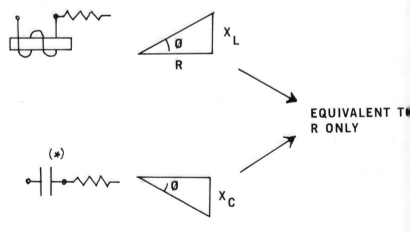

FIGURE III-15: Representation of PF correction

* Symbol for capacitance

44

Complete power factor correction of welding power supplies is not usually sought. Capacitors are quite expensive. Most of the time, the PF of a welding power supply is corrected to somewhere in the range of 0.8 to 0.9. The inductive PF angle is not quite cancelled out by a smaller capacitive PF angle.

Most welding power supply manufacturers offer PF correction as an option or a standard feature. Table III of the Appendix can be used to determine the required amount of capacitive voltamperes necessary for correcting poor PF in welding power supplies. The capacitors are usually connected across the incoming power leads to the power supply.

## Currents in Leading Circuits

The amount of current which flows in a capacitive circuit depends on the impedance of the capacitor. The impedance of capacitors is represented by the symbol $X_c$. It is measured in ohms. The symbol $X_c$ can be used in place of R in Ohm's Law to calculate current. As with inductors, do not use $X_c$ in Ohm's Law to calculate watts. Instead of watts, the capacitive circuit produces voltamperes, just as an inductive circuit produces voltamperes. However, capacitive voltamperes can be thought of as being the opposite of inductive voltamperes. They cancel each other when they are in a circuit together. The same rules of calculation are used for $X_c$ and for $X_L$, just remember that $X_c$ causes current lead and $X_L$ causes current lag.

The following equation determines the amount of $X_c$ in a 60 Hertz circuit.

III-e
$$X_c = \frac{1}{2\pi fC} = \frac{1}{377C}$$

Note that the C, for capacitance, is in the denominator. Capacitance is measured in Farads.

Example III-5:

**Find:** The current flow through a 0.001 Farad condensor when it is connected to an 80 volt 60 Hz supply.

**Solution:** From Equation III-e

$$X_c = \frac{1}{(377)(0.001)} = 2.65 \text{ ohms}$$

$$I = \frac{80 \text{ volts}}{2.65 \text{ ohms}} = 30 \text{ amperes}$$

**Answer:** 30 amperes

# CHAPTER IV

# THREE PHASE ALTERNATING CURRENT

## Introduction

A single sine wave of alternating current is called "single phase" current. Single phase current requires a circuit with three wires. Two wires are used to carry the current, and the third wire is used for a safety ground. The electrical needs are met by the two current carrying wires. The third wire is a necessary safety feature. In Chapter III, we discussed single phase AC. It is used mostly for low power requirements, such as in your home. Where large amounts of power are needed, such as in welding, three phase power is usually used to do the job. In addition, three phase power is more economical to use than single phase power. Less material is required in three phase equipment to do the same job as single phase equipment.

## What is Three Phase Power?

Three phase power refers to three sine waves of current flowing at the same time. Only one phase of current can flow in one pair of wires. It is not possible for two or more phases of current to flow in one pair of wires. Each single phase of current needs a separate pair of wires. Three phases of current could use as high as seven wires. Seven wires could be used to supply some type of electrical load as shown in Figure IV-1.

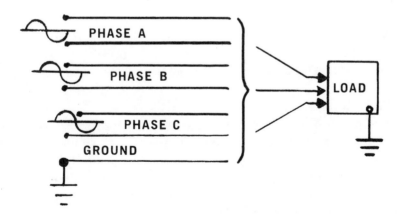

FIGURE IV-1: The Seven Wire, Three Phase Connection

For the sake of simplicity in power generation and in the construction of equipment, each of the three separate phases has a phase shift*of 120° from the other two phases. The 120° angle is used, instead of some other angle, because it is 1/3 of 360°. Remember 360° represents one full wave of a single phase of alternating current. Figure IV-2 shows this 120° relationship. The dashed lines are used to emphasize the phase shift in the sketch. If there was one more 120° phase shift for a 4th phase, the 4th phase would be in exactly the same place as phase A. Therefore, it wouldn't serve any purpose to have a 4th phase. It would only duplicate phase A.

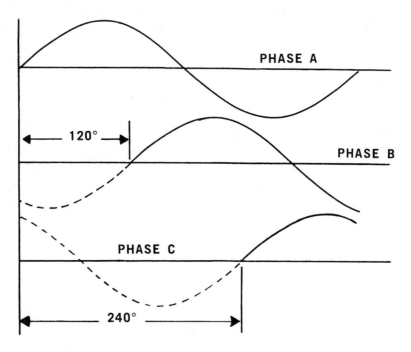

**FIGURE IV-2: The Phase Shift of Three Phase Power**

## The Four Wire System

It is possible to use 4 wires instead of 7 wires to carry three phase current to a welding system. Three wires are used to carry the current and the 4th wire is used as a safety ground. How is this possible, when in the previous paragraph it was stated that two wires are needed for each phase? The answer is simple. Each phase shares one, and only one, wire with another phase. That way, two currents never share the same two wires. This is shown in Figure IV-3.

* See phase shift, p. 31.

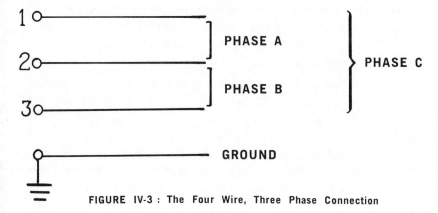

FIGURE IV-3 : The Four Wire, Three Phase Connection

Phase A shares wire 1 with phase C and wire 2 with phase B. Phase B shares wire 2 with phase A and wire 3 with phase C. Phase C shares wire 3 with phase B and wire 1 with phase A. Two wires are used for each combination, for a total of three wires and three phases. The fourth wire is a safety ground.

When any *one* of the three wires is removed, the circuit is broken for the *two* phases which share that wire. Only one phase is left complete. This is what happens when a fuse blows in one of the phases in a three phase welding system. When a three phase circuit has one of the three phases opened up, the circuit is said to be "single phasing." The four wire, three phase system is used for the obvious reasons of economy and simplicity. Four wires are easier to connect than seven wires.

### The Three Phase DELTA Connection

The voltages and currents in three phase systems are usually represented in diagrams by either a symbolic *delta* (triangle) or a

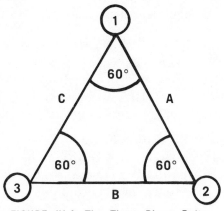

FIGURE IV-4: The Three Phase Delta

*wye.* Special rules are used by engineers to derive the diagrams, but it is not necessary for us to discuss those rules. The three phase *delta* (triangle) shown in Figure IV-4 summarizes many of the features of three phase power.

Each corner of the *delta* represents one terminal (wire) of a three phase system. The connection between any two corners represents one phase of the three possible connections. An open phase is represented by removing one of the corners. Removing one corner leaves only two corners or one phase left. The angle in each corner is 60°, two corner angles represent the 120° phase shift. The triangle can be used to represent current or voltage.

Many times you will see a triangular shaped circuit connection in a wiring diagram. The △ shape is used to emphasize a three phase circuit. Figure IV-5 shows a three phase* transformer connection.

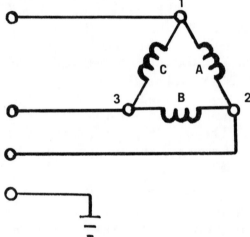

FIGURE IV-5: Delta Connection

The transformer windings are shown as the sides of the triangle. they are not usually arranged physically as a triangle. The windings are only shown that way in the diagrams. The triangular arrangement is called the DELTA connection. It is named after the Greek letter *delta* which looks like a triangle.

### The Three Phase WYE Connection

Another way of connecting a 3 Ø (three phase)* system is the WYE connection. It is called WYE because it looks like the letter Y. This connection is shown in Figure IV-6.

* The term "three phase" can be represented by the symbol 3 Ø

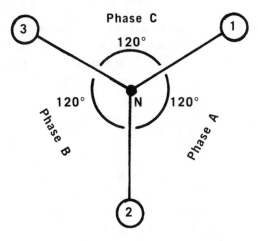

FIGURE IV-6: The Three Phase Wye

Just as in the delta diagram, the wye diagram has three angles. The angles represent the phase shifts. The numbers at the end of the wye represent the different terminals for the 3 Ø connections. They correspond to the terminals in Figure IV-3. The center of the wye is marked with an N for neutral. All of the currents and voltages cancel each other out as they flow through the neutral. This only happens at that point and no other. Since it is neutral, it is sometimes used as the ground connection. Figure IV-7 shows three transformers connected in wye. As with the delta connection, the wye connection is a Y only on paper. In a real circuit, the transformers are placed wherever it is convenient.

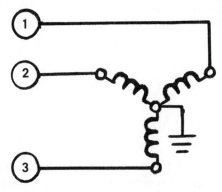

FIGURE IV-7: THE WYE CONNECTION

Three separate transformers can be connected either in Delta or Wye. However, the actual connection depends on the voltage rating of the transformer windings. In a Delta connection, each transformer receives the full input voltage from the power lines.

For example, in a 3 Ø , 440 volt AC △ system, each transformer would have 440 volts applied to its winding. In a wye connection, as you can see in Figure IV-7, there are two transformers connected between each pair of incoming power lines. The two windings share the phase voltage.

In a 440 volt wye connection, since the phase voltage is shared, each transformer operates at less than 440 volts. You might think each coil would have 220 volts because the two voltages add up to 440 volts. This does not happen because of the 120° phase angle. Actually, the voltage on each winding is 380 volts. The mathematics as to why it is 380 volts and not 220 volts is beyond the scope of this book, but it has to do with the phase shift of 120°. In Figure IV-8, the length of the triangle sides illustrates, after a fashion, how the 380 volts instead of 220 volts comes about. The sides of the triangle (1-N and 2-N) depend on the 120° angle. When the side between terminals 1 and 2 is 440 volts long, the other two sides are 380 volts long.*

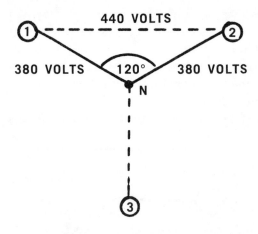

FIGURE IV-8: Three Phase Wye Voltages

Most of what has been discussed concerning 3 Ø circuits can be recalled by remembering the letters Delta and Wye. Each letter shape summarizes most of the rules related to 3 Ø systems. Just remember that 3 Ø systems are usually used where large amounts of power are needed. It is for this reason that most of high current welding systems use 3 Ø power.

* Refer to General Rule III-2, page 35

CHAPTER V

# ARC FUNDAMENTALS
# AND ELECTRICAL MODELS

## Introduction

All arcs have similar electrical characteristics. It does not make any difference to a power supply as to what kind of welding is being done. The volts and amps that the welding arc draws is all that the power supply cares about. Most of the time, an arc can be described by the simple model which is discussed in this chapter. The model does not explain all arc phenomena but it does give you a simple tool to figure out how power supplies and arcs work together in a system.

## The DC Arc Model

All arcs have an ANODE, CATHODE and PLASMA.* A detailed analysis of each of these parts is beyond the scope of this book. It is possible to model these parts of an arc by using batteries and resistors. The model is enclosed in a "dashed line" box to emphasize it is a model and not a power supply. The battery is used to represent the voltage of the anode and cathode portions. Resistors are used to represent the impedance of the plasma portion. Figure V-1 shows a simple model of an arc. It in no way represents the *true* arc characteristics, but it does represent the load on the power supply.

In the model:

$R_e$ = the Resistance of the electrode to which the arc is attached. It could be left out of the model and included with the model of the power supply. It is easier to put it here because in MIG and SUBMERGED ARC welding, it represents the resistance of the consumable electrode. It is shown as a variable resistor in the model because its value will change when welding conditions change.

$R_p$ = the Resistance of the plasma.

$E_x$ = the voltage (E) which is associated with changing arc length (x). It is shown as a variable battery because when the arc length is changed, this voltage changes.

$E_0$ = the voltage (E) which is associated with the anode and cathode. It doesn't change very much and is shown fixed in the model.

* Refer to Chapter I.

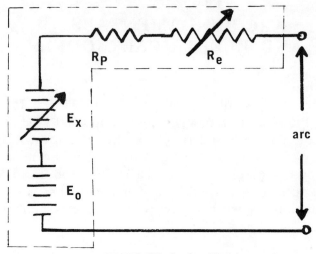

FIGURE V-1: An Arc Model

The parts of the model in Fig. V-1 can be combined to make the more simple model shown in Figure V-2.

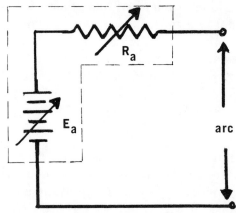

FIGURE V-2: A Simplified Arc Model

In this model:

$R_a$ = the variable *R*esistance of the *a*rc. It changes its value automatically to whatever the model needs. It represents the sum of $R_p$ & $R_e$ in Figure V-1.

$E_a$ = the variable voltage (E) of the *a*rc. It changes its value to automatically match that of an arc. It represents the sum of $E_O$ and $E_x$ in Figure V-1.

For example, imagine a model as in Figure V-2, with $R_a$ fixed at 0.01 ohms and $E_a = 0$. Then the model represents a simple fixed resistor. Its voltage and current characteristic* can be represented on a graph as shown in Figure V-3.

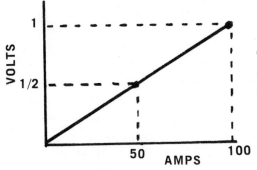

This line represents all the volt and ampere combinations for $R_a = 0.01$ ohms

FIGURE V-3: The Volt/Ampere Curve of a Resistor

In. Fig. V-3, at 100 amperes the 0.01 ohm resistor has a 1 volt drop (voltage drop $= IR = 100 \times 0.01 = 1$ volt).
At 50 amperes, the voltage drop is 1/2 volts ($50 \times 0.01 = 0.5$ volts). At 25 amperes, the drop is 1/4 volts, etc., etc., etc. For every combination of volts and amps, there is a point on the graph in Fig. V-3. The points form a straight line as shown. The line runs through zero volts at zero amperes.

Now imagine the same model with $E_a = 10$ volts instead of zero volts. At zero amperes, the arc terminals will measure 10 volts. When current flows through $R_a = 0.01$ ohms, the voltage rises above 10 volts. This is shown in Fig. V-4. The line starts at zero amperes and 10 volts, instead of zero amps and zero volts as in Fig. V-3.

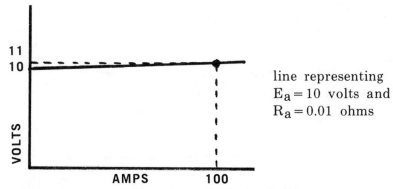

line representing $E_a = 10$ volts and $R_a = 0.01$ ohms

FIGURE V-4: The V/A Curve of a Simple Arc Model

* Also called the V/A curve.

Take the model one step further and vary the battery voltage ($E_a$) up and down, between 10 volts and 15 volts. This simulates a change in arc length made by the welder. The curve for the 10 volt arc model is shown in Figure V-4. The curve for the 15 volt arc model is shown in Figure V-5.

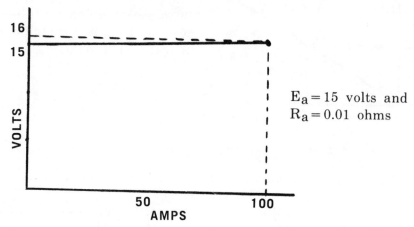

FIGURE V-5: The V/A Curve of a Simple Arc Model

The two curves shown in Figure V-4 and V-5 represent the minimum and maximum arc length characteristics for this particular model. Changes in the value of the resistor $R_a$ will change the *slant* of the V/A curve. Changes in the value of $E_a$ will change the *height* of the curve above the current axis.

When $R_a$ and $E_a$ are adjusted at the same time, it is possible to duplicate an actual arc as shown in Figure V-6.

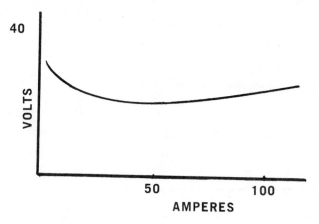

FIGURE V-6: The V/A Curve of an Arc

Every point on the curve represents a combination of volts and amps which an arc could use. The arc model represented by Figure V-2 and the curve in Figure V-6 is a good one, but not perfect. It is necessary to make a simple change in the battery-resistor arc model to make it almost perfect. When the arc model is shown as a battery and resistor, as shown in Figure V-2, the battery theoretically could deliver current. This defect of the model can be corrected by adding a rectifier.*

The complete model is shown in Figure V-7.

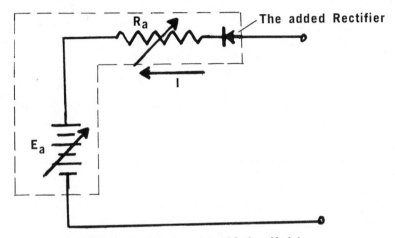

FIGURE V-7: The Complete DC Arc Model

The "arrow head" of the rectifier symbol shows the direction of allowable current flow. The "bar" represents a *bar*rier to the current flow. Current can only flow in the direction shown. The battery in the model cannot cause current to flow. The rectifier prevents the current flow. The arc model is complete.

### The AC Arc Model

Alternating current arc models are just like DC arc models, except they have two halves. One for each half cycle of AC. The two model halves are different. In one model, the arc current flows from a point (electrode) to a plane (workpiece); and in the other, the arc current flows from a plane to a point. The change in current direction is one reason why the two halves are different. In a sense, an AC arc is like two different DC arcs, switched in and out, alternately. An AC arc behaves like a reverse polarity DC arc

---

* A rectifier is like a "one way" valve, it lets current flow in one direction only. See Chapter VI, page 63.

for one half cycle then a straight polarity DC arc for the other half cycle. Figure V-8 shows the volt/ampere curve and arc model for a typical RP arc.

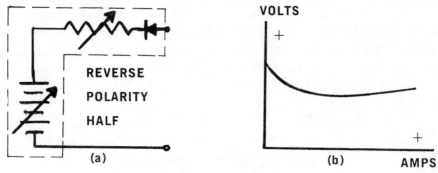

FIGURE V-8: A Typical RP Arc Model

Figure V-9 below shows a V/A curve and model for a typical SP arc.

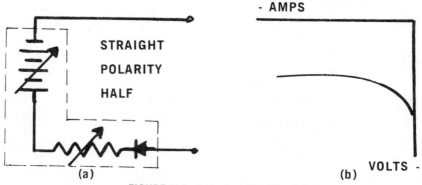

FIGURE V-9: A Typical SP Arc Model

The V/A curve in Figure V-9 is shown upside down because an SP arc is opposite in polarity to an RP arc. Since the RP arc was shown with positive polarity the SP circuit model is shown with the battery and rectifier reversed. The current will flow counterclockwise in the RP model circuit, and clockwise in the SP model.

The two halves shown in Figures V-8 and V-9 can be joined into one complete model by use of an imaginary, fast acting switch, operating 120 times a second. The switch operates alternately, sixty times in the RP direction and sixty times in the SP direction.

The joint volt ampere curves and models are shown in Figure

V-10. Actual experimental circuit models have been built, just as is shown in Figure V-10, and they have worked perfectly.

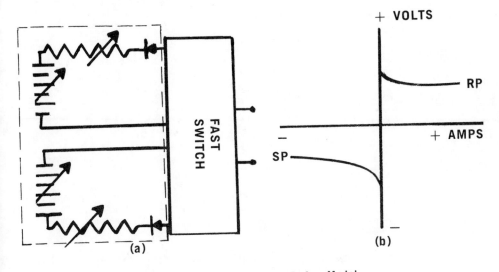

FIGURE V-10: The Complete AC Arc Model

## Summary

Arcs are very complicated electrical devices. However, they can be thought of in very simple electrical terms. Their performance can be analyzed by way of models using only batteries, resistors, rectifiers; and in the case of AC arcs, a switch as well.

Specific types of arc and their operation will be discussed in the chapters of Part II.

# PART II

# CHAPTER VI

# POWER SUPPLY
# CONSTRUCTION FUNDAMENTALS

## Introduction

Welding power supplies are made in many different ways, by different manufacturers, for a wide variety of welding processes. But, they all behave according to the same rules of electricity and they all have to do the same job—that is, operate a welding arc. This chapter discusses some of the basic electrical parts which most manufacturers use in building power supplies. The arc performance of the power supplies is discussed in other chapters of Part II; this chapter deals with power supply parts. It discusses the principles of operation of the parts which make up a power supply, *not* how they work in a welding system. Each sub topic of this chapter discusses a specific power source component, such as generators, rectifiers, contactors, etc.

## Generators

Generators use *motion, magnetism* and *wires* to produce electricity.* The motion in welding generator systems is provided by an engine, or an electric motor. The magnetism is usually produced by current flowing in stationary coils wrapped around iron cores, called pole pieces. Sometimes, permanent magnets are used. Most of the time, the current producing wires are wound on a rotating part, called the armature. The current is drawn from the generator armature by means of brushes running on a commutator. In some generators, the magnetic field revolves and the current producing wires are stationary.

Generators can be designed to produce either direct current or alternating current. The word ALTERNATOR is used when referring to AC generators. Most people have some knowledge of generators and alternators because they are the main source of electrical power in automobiles. Except for a difference in size and output, welding generators are much the same as automobile generators. The welding voltage is controlled by a voltage regulator unit, in much the same way as on an automobile. Generators are not discussed any further in this book. Any number of references are readily available for those who wish to know more about the operation of generators.

The theories of operation of welding systems, discussed in the

---

* Refer to Chapter I.

chapters of Part II, apply to generators just as well as to stationary power supplies.

## Rectifiers

Rectifiers are devices which allow current flow in only one direction. They block current in the other direction. They are "one way" streets for electrons. A rectifier can be represented by the symbol in Figure VI-1.

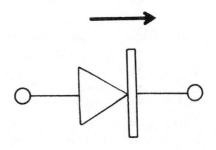

FIGURE VI-1: Symbol for a Diode*

The arrow like shape of the symbol indicates the direction of the current flow. Two rectifiers connected "back to back" or "head to head", as in Figure VI-2, completely block current flow.

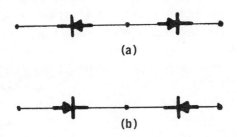

(a)

(b)

FIGURE VI-2: Diodes Connected to Block Current Flow

Rectifiers are made from several types of materials. Their shapes are indirectly determined by the material from which they are made. There are solid state rectifiers, gas and vapor filled glass envelope, vacuum tube, and mercury pool rectifiers, to name a few.

Vacuum tube rectifiers are used in TV sets, radios, etc. Their current rating is relatively low. They are not used in welding power supplies to any great extent.

Gas and vapor filled rectifiers are used in circuits where moderate power is required, such as in motor drives. Thyratrons are one popular form of this type of rectifier. They are used in some welding controls.

Mercury pool rectifiers have very high current capacity. They

* Diode is another name for simple rectifier.

are used where high power is needed, such as in electric locomotives. Ignitrons are a popular form of this type of rectifer. They are used in resistance welding equipment.

Solid state rectifiers are used to make most welding power supplies. A single solid state rectifier is called a DIODE. Diodes are made from the materials shown below:

- a. copper oxide
- b. selenium
- c. germanium
- d. silicon

The copper oxide rectifier is used in instruments. Selenium rectifiers are widely used in welding equipment. They are being replaced with silicon rectifiers in many cases.

Germanium rectifiers are not used very often in welding equipment, because silicon rectifiers are less expensive and are a little more rugged. Silicon rectifiers are used by most welding equipment manufacturers. Usually, the term "SOLID STATE" means the equipment is constructed without the use of glass containers or their equivalent. Silicon diodes, silicon controlled rectifiers (SCR's), selenium, germanium and copper oxide rectifiers are "solid". They are not made with glass envelopes. Selenium and silicon as shown in Figure VI-3, are the two most popular rectifier types used in welding power supplies. They can be readily identified from their shapes. The selenium type generally looks like a series of cooling fins on a shaft. They look quite like a steam pipe radiator with fins several inches square. Each radiator fin is part of one diode unit. Diode sections are added in series to increase the rectifier voltage rating. Silicon diodes look like a big bolt and nut with a braided copper pig tail. Silicon rectifiers are usually bolted

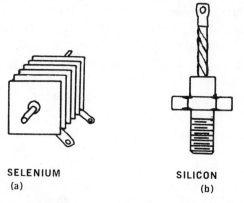

SELENIUM
(a)

SILICON
(b)

FIGURE VI-3: Sample Diode Assemblies

onto aluminum cooling plates or some special shaped cooling devices with fins. They have a much higher voltage rating than selenium units. When the voltage rating of a diode is exceeded, it will breakdown. Silicon diodes fail catastrophically. Selenium diodes sometimes "heal" themselves. In fact, on occasion, selenium diodes are connected across silicon diodes to protect the silicon units from voltage transient damage.

It is not possible to repair a selenium rectifier assembly. When it fails, the entire unit must be replaced. When a silicon diode fails, it can be removed, just as a bolt can be removed, and then replaced. It is not necessary to discard the cooling plates, etc. Only the damaged silicon diode need be replaced. The main application of rectifiers is to make direct current from alternating current. Rectification will be discussed in Chapter VIII.

### Contactors, Relays and Switches

Welding circuits, control circuits—all kinds of circuits—have to be turned *on* and *off.* The easiest way to do this is by opening a set of contacts. Contactors, relays and switches all use contacts to interrupt a circuit. Generally, the term CONTACTOR is used to describe a heavy duty relay which interrupts the main line power. The term RELAY is usually reserved for devices with contacts that are operated by an electro-mechanical mechanism. The term SWITCH is usually used for a device with contacts that are hand-operated. All three terms are apt to be used for each other. Such as saying "a *contactor* is a *relay* that *switches.*"

The contact material is one of the most important parts of a switch system. When a circuit is interrupted, the current wants to keep flowing.* Some electrons continue to jump the space between the contacts as they are separated. When the space between the contacts is made large enough, the electrons cannot jump the gap. Current flow stops. The circuit is turned off. The higher the voltage, the greater the gap which the electrons can jump. This is why low voltage switch devices should not be used in high voltage circuits. With high voltage, the electrons will continue to jump the contact gap. Even though the contacts are opened, the circuit won't turn off. Not only that, the heat generated by the arc between the contacts will destroy the contacts themselves and may cause a fire. In every sense, a sustained arc between contacts is just as hot as a welding arc.

The design of switching systems is complicated and requires a

* See General Rule II-6, page 27.

great deal of know how. Selection of the correct switch, contactor or relay is not as easy as it may seem. In addition to all of the electrical problems, there are problems of mechanical life, safety factors, mounting, cost, etc. Figure VI-4 shows two simple types of relays. When the coil of a relay is energized, the magnetic field of the coil causes the contacts to be pulled open or closed.

Relay with normally open contacts. Energize the coil to close the contacts.
(a)

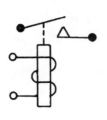

Relay with normally closed contacts. Energize the coil to open the contacts.
(b)

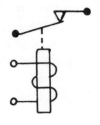

FIGURE VI-4: Example of Relays

## Fuses, Circuit Breakers and Thermal Cutouts

Electrical equipment should be protected from overloads. Overload of power supplies, motors, etc., will cause an increase in their temperature because of extra $I^2R$ heat. Unless the overload is removed, serious damage will occur to the equipment. Electrical overloads are accompanied by higher than normal currents. The higher current can be used to melt a fuse or trip a circuit breaker. Temperature sensitive devices can be used to operate relays or contactors to disconnect the load.

Most safety devices used in welding power supplies are based on the effects of $I^2R$ heat. A fuse is nothing more than a conductor with a calibrated resistance and low melting point. Normal current flow cannot create enough $I^2R$ heat to melt the fuse, but extra current can cause the fuse to "blow." A slight overload takes much longer to melt a fuse than a severe overload. Catastrophic overloads cause a fuse to melt (blow) immediately. The current-time relationship which causes a fuse to operate is dependent on the size of the fuse and the material from which it is made. Fuses have an $I^2t$ rating. The $I^2$ part of the rating comes from the $I^2R$ heat, and the $t$ comes from the time it takes the current to melt the fuse. The $R$ is the calibrated fuse resistance. The higher the current, the shorter the time it takes to blow a fuse.

Most circuit breakers are "tripped" by the $I^2R$ heat generated in a bimetal strip (two thin pieces of different metal sandwiched together). Current is passed through the bimetal strip. The $I^2R$

66

heat developed in the strip causes the metal strip to expand. The two different metals expand at different rates. The different expansions cause the bimetal strip bend and apply pressure to a tripping mechanism. Just as with fuses, circuit breakers have $I^2t$ ratings. The higher the current, the shorter it takes to open the circuit. Whereas fuses must be replaced when they are blown, a circuit breaker can be reset. Some circuit breakers use a magnetic field to trip the mechanism. Whether it is heat or magnetism which trips the breaker, the result is the same—the circuit is interrupted and protected.

Thermal cut offs are nothing more than bimetal strips arranged to operate small contacts. The mechanism is usually sealed in a small, electrically insulated tube with the switch leads brought out. The insulated tube can be placed inside a current carrying coil, in touch with a transformer core, on a rectifier cooling plate, or wherever a current overload will cause a temperature rise. When the thermal cutoff switch contacts close, they can be used to operate a relay or contactor to interrupt the main power. They are available in a wide variety of shapes and sizes, over a wide range of temperatures.

Some power supplies are designed with these protective devices as standard equipment, while others have them offered as optional equipment. Unless some type of safety device is used, chances are that the power supply will be severely damaged by long term overloads. Continued operation at moderate overloads will decrease the life of electrical equipment. Overload of electrical equipment is to be avoided.

**Meters**

Ammeters and voltmeters are the two most common meters found in welding systems. The vast majority of these meters work in the same way. They all depend on the universal concept of *motion, magnets* and *wires* as was discussed in Chapter I and in the section on Generators in this chapter. A very small portion of electrical power is taken from the system being measured, and is used to move the pointer of the meter. The meters are very sensitive. They do not use any noticeable power and do not affect the welding arc.

Most meters consist of a small coil of wire wrapped around a moveable core, to which a pointer has been attached. The small coil core is pivoted so it is free to move. A very weak spring connects to the core to hold it in place. The coil is placed in the field of a permanent magnet. When a small current flows in the coil, it

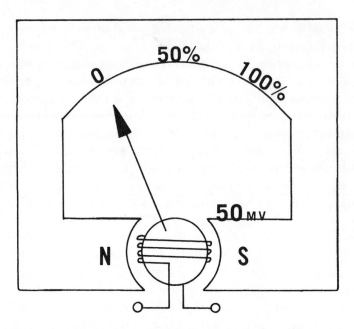

FIGURE VI-5: Simple Meter Movement

magnetizes the coil core. The magnetic field of the coil pushes against the field of the permanent magnet. The greater the coil current, the greater the rotation of the core on its pivots. The pointer is used to show how far the coil rotates. The spring returns the pointer to zero when the current flow stops. It is a simple matter to calibrate the meter scale to show how far the pointer moves with a specific amount of meter current. A simple meter is shown in Figure VI-5.

Many meter movements are designed to give full scale deflection when 50 millivolts is applied to the meter coil.

An ammeter is nothing more than one of the 50 millivolt meter movements connected across a resistor, as shown in Figure VI-6. The resistor is selected to have a 50 millivolt drop across it when a specified load current flows through it. The resistors are called Shunts. A 500 ampere shunt will produce a 50 millivolt drop when 500 amperes flow through it. A 10 ampere shunt will produce a 50 millivolt drop with 10 amperes. A 50 mv* meter movement can be connected across any 50 mv shunt and give a full scale reading when full shunt current is flowing. Manufacturers of ammeters match the meter scale with the shunt value, but the basic meter movement is the same. Any 50 mv meter movement can be used with any 50 mv shunt, it is just a matter of con-

*mv = the abbreviation for millivolt.

venience to use a meter whose scale is matched to the shunt current rating.

Voltmeters are nothing more than meter movements with current limiting resistors connected in series with the coil as shown in Figure VI-7. The coil current produces full scale deflection when the meter circuit is connected to a specific voltage. The resistors, which are connected in series with the meter movement are used to keep the current in the meter coil at a safe level. Voltmeters are usually listed in catalogues as having 20,000 ohms per volt, or 100,000 ohms per volt, or some such number. For example, a

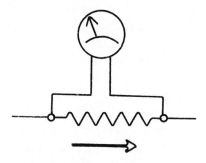

FIGURE VI-6: A Simple Ammeter

20,000 ohm per volt meter with a 100 volt scale would have a (100 x 20,000) 2,000,000 ohm resistor in series with the coil.

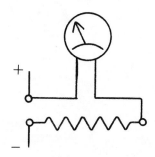

FIGURE VI-7: A Simple Voltmeter

Meter costs depend on a variety of factors such as ruggedness (ability to stand mechanical shocks), accuracy, repeatability, sensitivity, etc.

Chart recorders and a variety of other instruments work on variations of the *motion, magnetism* and *wire* principle. The 50 mv meter movement is not the only type of measuring device available, it just happens to be the most popular. It is in use in many places, not just welding power supplies.

**Inductors and Reactors**

An inductor is nothing more than a coil wound on a core,* its main purpose is to control the rate of change of current in a circuit. It does this by temporarily storing electrical energy in the form of a magnetic field. Then when the field collapses, it returns the stored energy to the circuit as voltage. Several types of inductors are shown in Figure VI-8. As a matter of convenience, an *inductance* coil is usually called an *inductor* when it is used in *DC* circuits. When an inductance coil is placed in an *AC* circuit, it is usually called a reactor.

Inductors used in welding circuits are usually quite large. They can weigh up to several hundreds of pounds or as little as 25-30 pounds depending on the current rating and amount of inductance. Several types are shown in Figure VI-8.

At various times, the following terms have been used for inductance coils in direct current welding circuits:

| INDUCTOR | STABILIZER |
|---|---|
| REACTOR | CHOKE |
| DYNAMIC REACTOR | FILTER CHOKE |
| WELD STABILIZER | |

**Air Core Coil**
**needs many turns**

**Open loop iron core**
**needs less turns**

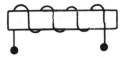

**Closed loop iron core**
**needs least turns**

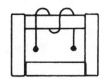

**FIGURE VI-8**

70

* It obeys the laws of *motion, magnetism,* and *wires* discussed in Chapter I and Chapter II.

## Transformers

Transformers use the principle of *motion, magnetism* and *wire* to change alternating current from one level to another. There are voltage transformers and current transformers. As discussed in Chapter I, the *motion* of a transformer is supplied in the form of a fluctuating (alternating) *magnetic* field. The *wires* (coils) are stationary. A simple transformer is shown in Figure VI-9. The coil voltage depends on the number of coil turns. More turns give more voltage.

It is possible to use one of the coils shown in Figure VI-9 to produce the fluctuating magnetic field, by connecting it to a source

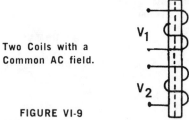

Two Coils with a
Common AC field.

**FIGURE VI-9**

of power. This is shown in Figure VI-10. The magnetic field produced by the first coil (primary) passes through the second coil (secondary) and generates a voltage in the second coil. When the two coils have the same number of turns, the voltage generated in the second coil is the same as the voltage applied to the first coil. When the second coil has less turns the voltage is less, with more turns the voltage is greater. Figure VI-11 shows two types of transformer illustrations.

Two coils with a common
field produced by one
of the coils

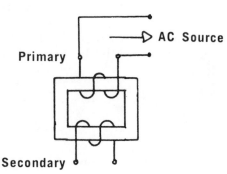

**FIGURE VI-10**

71

As shown in Figure VI-12, transformer coils may be connected in series with each other to increase voltage. When they are connected in series, but with their terminals reversed, the voltages of each coil are subtracted, instead of added.

(a)

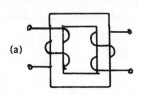

Pictoral Representation of
a two coil transformer

(b)

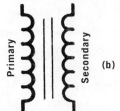

Symbolic Representation of a
Two Coil Transformer

**FIGURE VI-11**

Transformer with two series connected
secondary coils

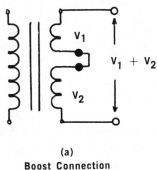

(a)
Boost Connection

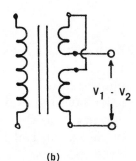

(b)
Buck Connection

**FIGURE VI-12**

As shown in Figure VI-13, the transformer coils may also be connected in parallel to increase the current output at a particular voltage.

Transformer With Two Parallel
Connected Secondary Coils

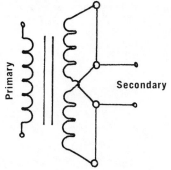

**FIGURE VI-13**

There is even one form of transformer where the secondary and primary are connected together, to make what looks like a single coil transformer. This is called an AUTOTRANSFORMER. When the single coil is wound on a doughnut shaped core with its output connected to a sliding brush, as shown in Figure VI-14, it is called an adjustable autotransformer. It is marketed under various trademarks such as: POWERSTAT, VARIAC, VOLT-PAC, OHMITRAN

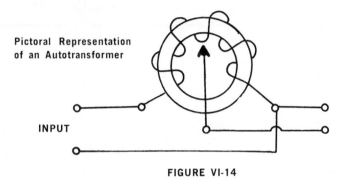

Pictoral Representation
of an Autotransformer

INPUT

**FIGURE VI-14**

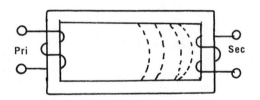

Transformer with a long core
and leakage flux

Pri

Sec

**FIGURE VI-15**

Whenever a transformer is connected to a load, the current in the transformer coils tends to "squeeze" some magnetic flux outside of the iron core. The magnetic flux is said to "leak" around the secondary coil, as shown in Figure VI-15. As more and more flux leaks around the secondary coil, the secondary voltage becomes smaller and smaller. When all the flux is caused by "leak" around a coil, by special transformer core designs, the voltage becomes zero. With no magnetic flux, there is no voltage.

Many welding power supplies use transformers of special mechanical design. The special designs are used to control the amount of leakage magnetic flux. One such design with a flux bypass is shown in Figure VI-16.

73

Transformer with a magnetic
shunt for flux by-pass

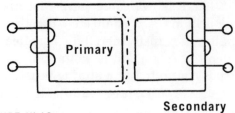

Primary

Secondary

FIGURE VI-16

Several other mechanical designs are illustrated in Figures VI-17 through VI-20. Each of the designs follows the laws relating to *motion, magnetism* and *wire.*

**Basic Three**

**Legged Core**

- - - -

**No adjustment**

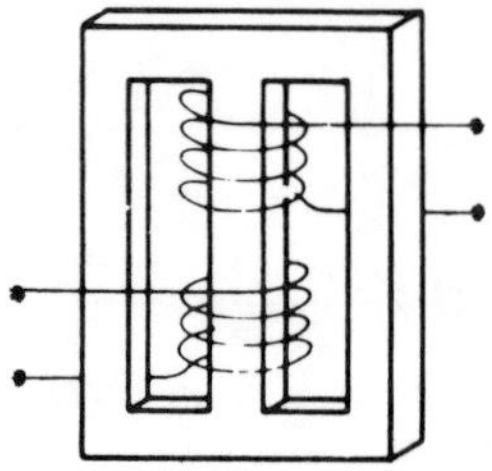

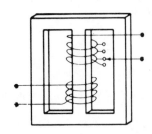

FIGURE VI-17

**Three Legged**

**Core**

- - - -

**Tapped Secondary**

**Changes output**

FIGURE VI-18

**Three Legged Core, move-**

**able magnetic shunt**

- - - -

**Leakage flux changes**

**with shunt position**

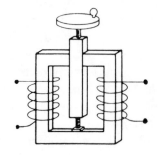

FIGURE VI-19

Three Legged Core, move-
able coil adjustment

- - - -

Leakage flux changes
with coil position

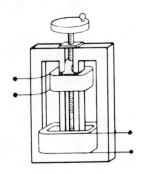

FIGURE VI-20

# CHAPTER VII

# SOME WELDING POWER SUPPLY FUNDAMENTALS

## Introduction

Welding power supplies come in all sizes and shapes; they may supply AC or DC or both; they can be bought with all types of features and optional equipment. However, unless the power supply can supply the correct power for the welding arc, it is useless. Different welding processes need different types of power. A power supply designed for use with covered electrodes (stick) will be less than satisfactory for consumable electrode (MIG) welding systems.

Most welding power supplies have several features in common. This chapter discusses those features. Later chapters cover the specific power supply requirements for specific arc welding processes.

## Volt/Ampere Curves—Power Supply Slope

The output characteristics of a power supply can be represented on a graph of volts vs. amperes. (This is the same as was done in Chapter V, for arc characteristics.)

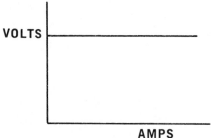

FIGURE VII-1: Idealized Volt/Ampere Curve of a Transformer or Battery

The volt/ampere curve (VA curve) of an ideal transformer, or battery can be represented by a straight line, as shown in Figure VII-1. The line shows that no matter what current is drawn from the transformer, or battery, the voltage remains constant. Theoretically, there is no limit to the amount of current which could be drawn. The same curve even represents the output of the Public Utility which supplies the AC to your home. No matter how many appliances are used (increased current draw) the voltage remains constant.

*Real* transformers and batteries do not provide constant output voltage at all currents. Real power supplies are made with wires and materials that have resistance and/or some impedance. It takes voltage to push current. The voltage at the terminals of the power supply is reduced when current flows through its internal resistance. A similar reduction in voltage occurs with the power supplied by a public utility. Too much load can cause local "brown outs" because of the voltage reduction. For example, if a battery had an internal resistance of 0.02 ohms, the terminal voltage would drop off 2 volts for every 100 amperes drawn from the battery. This is shown in Figure VII-2.

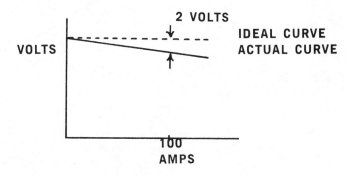

FIGURE VII-2: Volt Ampere Curve of Power Supply with 0.02 Ohms Internal Resistance.

The output voltage of the battery will be at some proportional value at other currents. In power transformer technology, the *fall off* of voltage with an *increase* in current is referred to by the term "regulation." The regulation of a transformer is its *no load terminal voltage* minus its *load voltage* divided by the *load voltage* and is expressed in percent.

VII-a % Regulation = $\dfrac{\text{NO LOAD VOLTAGE - LOAD VOLTAGE}}{\text{LOAD VOLTAGE}} \times 100$

In welding, the term SLOPE is used to describe the regulation. Instead of giving the percent regulation for a welding power supply, most welding manufacturers give the Slope. The slope is expressed in volts drop per 100 amperes of current flow. In Figure

VII-2, the slope of the line is 2 volts/100 amperes.

The term slope is derived from the downhill slant of the V/A* curve. The slope of a hill is its *rise* divided by its *run*. Power supply slope is defined in the Figure VII-3.

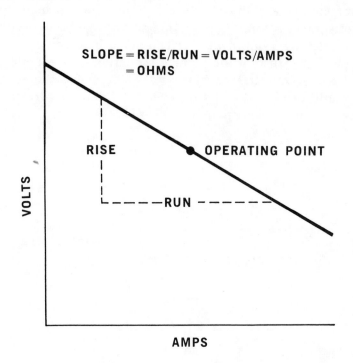

FIGURE VII-3: Definition of slope.

A power supply with a great deal of internal resistance, or impedance, will have a "steep" slope. A power supply with very little internal loss will have a "flat" slope.

A flat slope power supply is also referred to as a *constant voltage* or constant *potential* unit. Its output voltage remains essentially constant over the normal operating range of currents. Such power supplies are sometimes called CP or CV units.

Some power supplies have a very steep slope. These units are sometimes referred to as "droopers" because the droop (slope) of the V/A curve is very steep. Their output current is essentially constant in the normal operating range. A typical "drooper" power supply V/A curve is shown in Figure VII-4.

* V/A is the abbreviation of volt/ampere

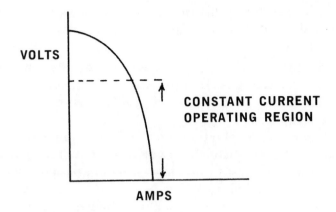

VOLTS

CONSTANT CURRENT
OPERATING REGION

AMPS

FIGURE VII-4: V/A Curve of Typical Constant Current Power Supply

Power supplies with steep slopes are used for a number of welding processes, typically for covered electrode and TIG welding. These power supplies are also called "conventional" or "constant current" units. The term "conventional" refers to the fact that it is the usual, or customary, power supply used for covered electrode welding. Such power supplies are sometimes referred to as CC units.

It is possible to build power supplies with all types of slopes by use of feedback circuits, magnetic amplifiers, solid state circuits, etc. One very useful MIG welding power supply combines constant

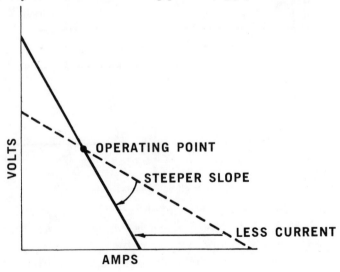

VOLTS

OPERATING POINT

STEEPER SLOPE

LESS CURRENT

AMPS

FIGURE VII-5: Effect of changing slope

potential output with a moderate amount of slope adjustment. It is usually called a "slope controlled CP" power supply, or a "CP controlled slope" unit.

The specific use of slope in a *welding* power supply will be discussed in later chapters. It is sufficient at this point to mention that slope has a very definite, and a very important, effect on the welding performance.

When used in combination with a voltage adjustment, it is possible to use slope adjustment to vary the slant of the power supply V/A curve as it passes through a specific arc operating point. This is shown in the Figure VII-5. Note that the no load voltage (open circuit voltage) of the solid line is higher than that of the dashed line, and its slope is greater (steeper).

The use of voltage and slope in various combinations can generate families of power supply output curves such as are shown in Figures VII-6 thru VII-8. The figures are representative of typical output volt/ampere curves of commercial power supplies.

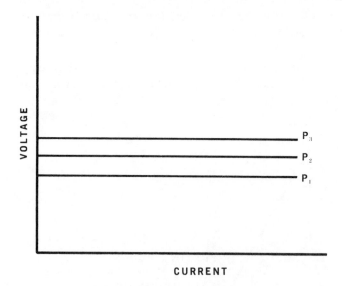

CURRENT

FIGURE VII-6: Constant Potential

Slope can be provided by any electrical device that has an impedance. In fact, as far as the arc is concerned, anything which adds impedance adds slope: power cables, connections, loose terminals, dirty contacts, transformer windings, rectifier losses, etc. Slope adjustment can be obtained by use of a tapped resistor connected in series with the output leads of the power supply. In AC

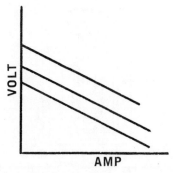

FIGURE VII-7: CP with Slope

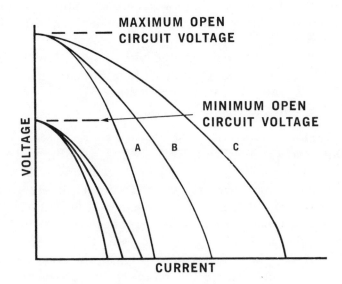

MAXIMUM OPEN
CIRCUIT VOLTAGE

MINIMUM OPEN
CIRCUIT VOLTAGE

A    B        C

VOLTAGE

CURRENT

FIGURE VII-8: Conventional (Droop) Type

power supplies, it is also possible to use a tapped reactor connected in series with the output leads. Other types of slope reactors include moving coil, moving core, magnetic amplifiers and other devices. Some of these slope devices are shown in Figure VII-9.

81

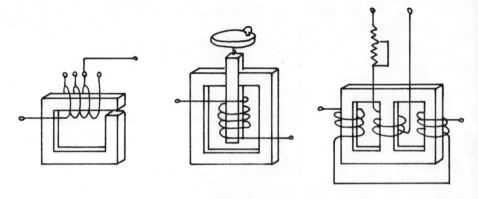

FIGURE VII-9: Typical Slope Reactors

The tapped adjustment can be provided by switches, sliding brushes on a commutator, terminals with links, etc. There are any number of ways to build slope* into a power supply.

Constant potential Mig power supplies with slope adjustment have been sold under some of the following commercial descriptions:

a) VS units, for *voltage* and *slope*, or *variable slope*

b) SV units, for *slope* and *voltage* adjustments

c) RAV units, for *rising arc voltage*

d) CS units, for *controlled slope*

### Duty Cycle

Duty cycle is another feature that all welding power supplies have in common. It is expressed in percent and tells how long a power supply can be run at its *rated* output. Unless a unit has a 100% duty cycle, it cannot be run continually. It must be turned off for a cool down period. Power supplies are not 100% efficient, only ideal power supplies have no losses. The $I^2R$ losses inside of real power supplies cause them to heat up. Without adequate cooling, it is not possible to operate 100% of the time.

A power supply that is overloaded can heat up to where damage will occur. Usually it takes some time for the heat to build up. The overload should be disconnected before damage occurs.

A 10 minute test period is usually used to determine the duty cycle of welding power supplies.** A power supply that can be

---

\* Don't confuse "slope" as used in Arc welding with "slope" as used in Resistance welding. In resistance welding, "slope" refers to current vs. time programming. *Up slope* and *down slope* mean increasing and decreasing currents. Some manufacturers of arc welding equipment manufacturing current vs. time control devices for programming the welding current. Occasionally, these too are called "slope controls."

\*\* Some countries use a 12 minute test period.

operated from one ten minute period to the next, without interruption, for unlimited time, is said to have a 100% duty cycle.

When a power supply must be turned off for cooling, in each ten minute test period, the duty cycle is less than 100%. For example, when rated load in *ON* for 8 minutes and *OFF* for 2 minutes, in a 10 minute cycle, the power supply is said to be operating at 80% duty cycle. A power supply with a 60% duty cycle can be operated at a rated load for 6 minutes (with a 4 minute cooling period) in one ten minute cycle.

Generally, the $I^2R$ heat is the limiting factor in determining duty cycle. Since the internal resistance (R) of a power supply is usually constant, it is sometimes possible to use current squared ($I^2$) ratios to determine the duty cycle at some other current. Less current means less heat, and this means the power supply can run for a longer time. For example, at 1/2 the current, the $I^2$ is 1/4 of what it would be normally. Operating for 4 times as long with 1/4 the $I^2R$ will produce the same total heat as when operated at full current.

The duty cycle of a power supply can be estimated at other currents by use of the following formula:

VII-b $$\% \text{ Duty Cycle} = \frac{(\text{Rated Current})^2}{(\text{Load Current})^2} \times (\text{Rated Duty Cycle})$$

CAUTION: The current rating of some individual components may be limited. Do not expect a power supply to operate at low duty cycles and high current unless the current is within the capacity (rating) of all the individual components (diodes, fuses, contactors, etc.)

The rated duty cycle of a power supply is determined by the manufacturer according to test standards published by NEMA.* Some manufacturers provide duty cycle curves similar to those shown in Figure VII-10. The curves are based on the duty cycle equation VII-b.

* National Electrical Manufacturers Association

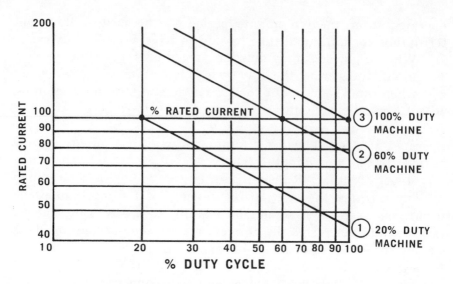

FIGURE VII-10: Universal Duty Cycle Curves

Figure VII-10 illustrates the curves for machines rated at 20, 60 and 100% duty cycles. The *% of rated current* is plotted against the *% duty cycle.*

Example VII-1:

> **Find:** the duty cycle of a machine rated at 100 amperes and 60% duty cycle, when it is operated at 150 amperes (a 150% load).
> **Solution:** From equation VII-b,
> $$\% \text{ Duty Cycle} = \frac{(100)^2 \times 60}{(150)^2}$$

**Answer:** 27% Duty Cycle

Example VII-2:

> **Find:** the output current of a 100 ampere 60% duty cycle supply, when it is operated on a 100% duty cycle.
> **Solution:** From a rearrangement of Equation VII-b,
> $$(\text{Load Current})^2 = \frac{100^2 \times 60}{100} = 6000$$

**Answer:** Load Current = 77 amperes, approximately

84

The result of Example VII-2 may also be found from the No. 2 curve in Figure VII-10. It shows 77% rated current at 100% duty. In this case, 77% of 100 amps is 77 amperes.

Similarly, Curve No. 1 in Figure VII-10 shows a power supply rated for a 20% duty cycle may be operated on a 55% duty cycle when the output current is reduced to 60% of the rated load. The same power supply can be operated on an 80% duty cycle when the output current is reduced to 50%.

Curve No. 3 shows that a machine with a 100% duty cycle may be operated at a 150 percent load if the duty cycle is reduced to 44% and at a 200 percent load if the duty cycle is cut to 25%, etc., etc., etc.

Although these curves may be used as a guide, care must be taken to avoid overloading power supply components even for brief period. Failure to do so may result in blown fuses or permanent damage to the power supply.

# CHAPTER VIII

# RECTIFICATION PRINCIPLES

### Introduction

Consumable electrode (MIG, SUBARC, FLUX CORE, ETC.) welding is usually done with DC power. Most of the power supplies are transformer-rectifier units. They convert AC into DC. The paragraphs that follow discuss the principles of rectification (changing AC into DC). These principles apply to all types of rectifiers. Whether they are selenium, copper oxide, silicon, etc., it makes no difference in their operating principles. The same principles apply for the smallest power supply (as in model railroads) to the largest (as in Traction motors of full size trains). Welding power supplies are just one of the many systems that use rectifiers to change AC to DC.

### Half Wave Rectification-Single Phase

Every sine wave of AC has two halves, a top and a bottom (plus and minus). It is possible to draw an imaginary switch circuit which can allow either half to pass and block the other. This is shown in Figure VIII-1:

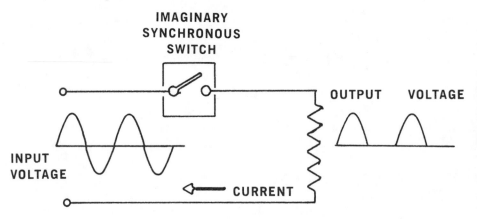

FIGURE VIII-1: Mechanical Half Wave Rectifier

The imaginary switch in Figure VIII-1 closes when the input sine wave is positive and opens when it is negative. When the switch is closed, the positive half of the input wave supplys current to the resistor (load). The switch opens every time the negative half wave comes along and stops current from flowing. A real switch could do the same thing, but it would probably wear out after a month or two of operation. The switch would have to open and close 60 times a second with 60 Hz power.

A diode will do the same job as the switch, but with no moving parts; and it will not wear out. The diode can run forever, as long as it is operated within its current and voltage rating. Figure VIII-2 shows how this is done.

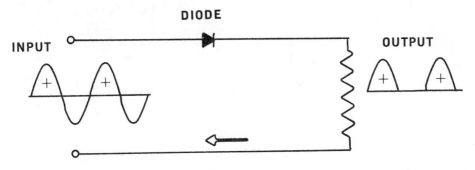

FIGURE VIII-2: Solid State Half Wave Rectifier

In Figure VIII-2, when the current flows in the same direction as the arrow of the diode symbol shows, the circuit acts as if the switch in Figure VIII-1 was closed. As soon as the sine wave goes negative, and the current tries to reverse itself, the circuit acts as if the switch was open.

The output current wave shape is called *half wave rectified DC,* or pulsating DC. By itself, it is not too useful for arc welding. Not only that, it throws away 1/2 of the available input wave.


## Full Wave Rectification—Single Phase

With full wave rectification, both halves of the input sine wave are used. Instead of blocking out 1/2 of the wave as in 1/2 wave rectification, full wave rectification uses the entire wave.

An imaginary switch circuit is shown in Figure VIII-3. Instead of one switch, this circuit uses four switches.

The switches operate in pairs A & C and B & D. When one pair is open, the other pair is closed, and vice versa. Assume that the A-C pair are closed when the input wave is positive, and that the B-D pair are closed when the wave is negative.

Figure VIII-4 shows the path of current in the circuit of Fig. VIII-3 when the input is positive. The current flows in a clockwise direction from the input.

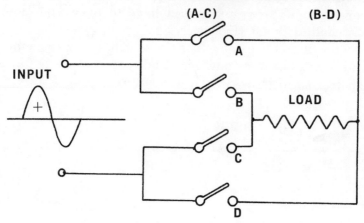

**IMAGINARY SYNCHRONOUS SWITCH PAIRS**

(A-C)  (B-D)

INPUT

LOAD

A

B

C

D

FIGURE VIII-3: Mechanical Full Wave Rectifier

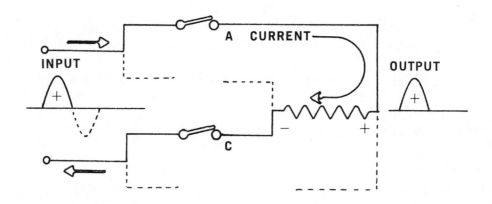

INPUT

A  CURRENT

OUTPUT

C

FIGURE VIII-4: Clockwise Current Flow Of Figure VIII-3, With Switch Combination A-C

As soon as the sine wave input reverses, A-C open up and B-D close. This is shown in Figure VIII-5:

Since the input wave is negative, the current is reversed, and is shown flowing in a counter-clockwise direction at the input terminals. There is one IMPORTANT feature which is common to Figures VIII-4 and VIII-5. The current in the load resistor is in the same direction. It has been routed by the switches to flow through the load in the same direction.

In a sense, the imaginary switches have taken the AC sine wave and *folded* the *negative halves up* between the positive halves. This is shown in Figure VIII-6.

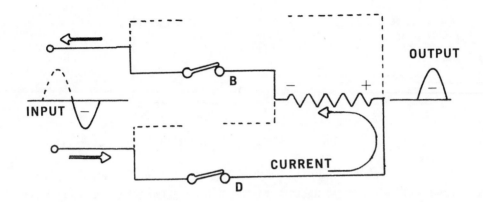

FIGURE VIII-5: Counter-Clockwise Current Flow of Figure VIII-3, With Switch Combination B-D

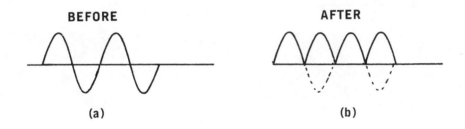

FIGURE VIII-6: Full Wave Rectification of a Single Phase Sine Wave

The imaginary switches of Figures VIII-3 thru VIII-5 can be replaced by diodes (rectifiers) as shown in Figure VIII-7. The diode circuit will produce the same rectification effect as the switch circuit.

In Figure VIII-7, when the input wave is positive and the current flows in a clockwise direction, it flows through diodes A and C. The clockwise current is blocked by diodes B and D. The opposite occurs when the wave is negative and the current flows in a counter-clockwise direction.

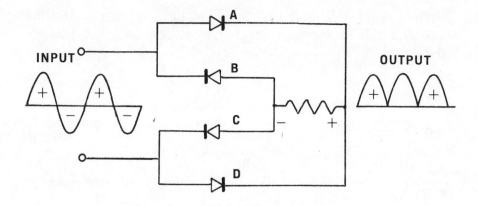

FIGURE VIII-7: Full Wave Bridge Rectifier

The circuit shown in Figure VIII-7 can be drawn in several forms. The most popular versions are shown in Figure VIII-8. A careful check of the wires and directions of current flow will show that the circuits in Figures VIII-7 and 8 are the same, even though they may look different.

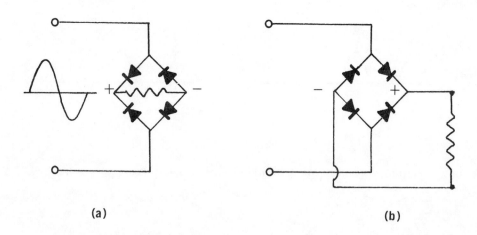

(a)                              (b)

FIGURE VIII-8: Typical Single Phase Full Wave Bridge Rectifier Circuits

The rectifier circuits shown in Figure VIII-8 are called *Full Wave Bridge* rectifiers. They do not have a "diamond" shape in the actual power supply circuit. Rectifiers usually come in complete assemblies and at first glance look very much like some sort of radiation fins, which in fact the fins (to which the diodes are attached) really are. A typical bridge assembly is shown in

Figure VIII-9. The diode elements that are used in the manufacture of a rectifier assembly are not perfect. They have internal resistance and get hot as a consequence of the current flow. The radiator fins aid in dissipating the heat which is generated by the current flow.

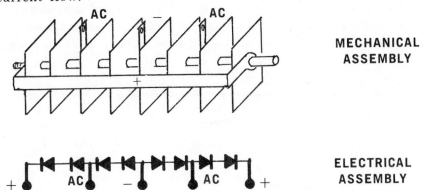

MECHANICAL
ASSEMBLY

ELECTRICAL
ASSEMBLY

FIGURE VIII-9: Typical Full Wave Bridge Rectifier

Manufacturers sometimes follow a color coding standard when building rectifier assemblies.* The line AC input terminals (the top and bottom corners of the "diamond" in Figure VIII-8) are color coded yellow; red is used for plus and black is used for minus.

When Silicon diode units are used to make a rectifier assembly, the individual units usually have a rectifier symbol marked on their sides. The units are assembled to make a circuit diagram just as is shown in Figure VIII-8.

### Rectifier Filters

The unfiltered output wave shape from either a half wave or full wave bridge rectifier with single phase input is not too useful for consumable electrode welding processes. There is too much waviness, or ripple, in the output. Every time a half wave of current goes to zero, the arc goes out and has to be reignited. One way to decrease the waviness of rectified single phase AC power is to filter the output wave shape. Filters are used to fill in the "valleys" between the pulses of current. They do this by storing up energy while the pulses are high and delivering the energy back when the pulses are low.

Generally, there are two main types of single phase full wave bridge** rectifier filters used in welding power supplies. One type is

*NEMA—ICS-1970
**1∅ FWB is a symbol commonly used for *Single Phase Full Wave Bridge.*

a pure inductive filter and the other is a combination of capacitance and inductance.

### Inductive Filters

This type of filter is nothing more than an inductor (choke) placed in series with the load circuit. A typical connection is shown in Figure VIII-10. The inductor could be placed in the negative lead instead of the positive lead as shown.

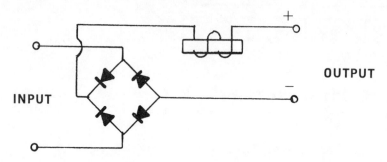

FIGURE VIII-10: Full Wave Bridge With Choke Filter

Any one of the many versions of an inductor can be used as a filter. The inductor stores energy in its magnetic field while current is flowing. In doing so, it decreases the waviness of the current passing through the inductor. When the current pulse starts down, toward a valley between pulses, the magnetic energy is released. The current flow is maintained during the valley. This is shown in Figure VIII-11.

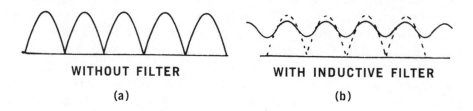

WITHOUT FILTER
(a)

WITH INDUCTIVE FILTER
(b)

FIGURE VIII-11: Typical Output Wave Shape of Single Phase Full Wave Bridge Rectifier

The 1Ø FWB (Single phase full wave bridge) rectifier output would be a straight line, with no ripple, if the inductor was infinitely large. An infinitely large inductor would oppose *all* change of current.

The bigger the filter inductor, the more it costs. Therefore, the cost of a power supply goes up when better filtering is provided. Design of power supply filters is a compromise between cost and filter performance.

### Inductive-Capacitive Filters

Inductors are not the only energy storage devices that can be used for filtering. Condensers also store energy. A capacitor can be connected to a 1Ø FWB rectifier as shown in Figure VIII-12

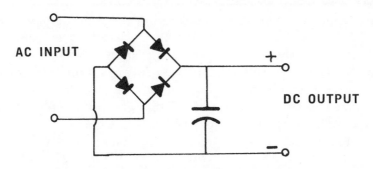

**FIGURE VIII-12: Full Wave Bridge Rectifier With Capacitive Filter**

The capacitor stores up energy (electrons) when a pulse is delivered from the rectifier to the load. The capacitor discharges the energy (electrons) to the load when the pulse from the rectifier heads toward zero. The charging and discharging of the capacitor tends

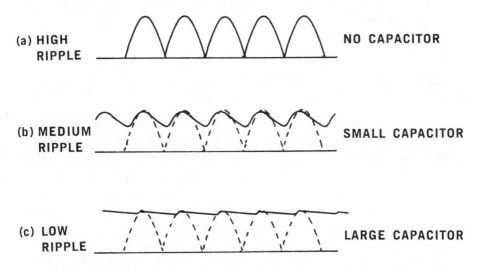

FIGURE VIII-13: Typical Output Waves of Capacitive Filter Single Phase full wave rectifier

to fill in the valley between pulses. A very large capacitor would fill the valley completely. There would be little or no ripple. This is shown in Figure VIII-13.

As with inductors, the cost of a capacitor filter varies with the amount of filtering desired. There is one other factor that should be considered in the use of capacitive filtering, and that is the capacitor charging current. Capacitance draws a charging current. The charging current is supplied by the rectifier. Unusually large amounts of capacitance could draw excessive current flow from the rectifiers and might damage them:

Pure capacitive filters are not used in arc welding power supplies. The capacitor in Figure VIII-12 will "hog" the current from the arc and cause the arc to go out, or be "snuffed". The same "arc snuffing" action of condensers is used in an automobile ignition system where a condenser is used across the "points". The condenser increases the operating life of the ignition points by "snuffing" the arc between the points when they are opened and closed.

Another version of filter circuits in welding power supplies uses both an inductor and a capacitor as shown in Figure VIII-14.

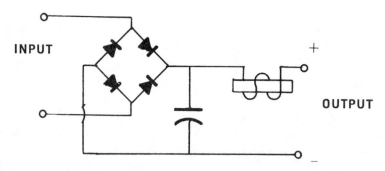

FIGURE VIII-14: Typical Single Phase Full Wave Bridge Rectifier with Inductor-Capacitor Filter

The inductor prevents the arc snuffing action of the capacitor from extinguishing the arc. The combination provides good filtering at a moderate cost. The capacitor provides good filtering at low currents and the inductor at high currents. The magnetic field of the inductor is not very great at small currents. Consequently, its filtering action is poor at low current. With high arc currents, the capacitor is required to discharge completely during the "valley" of the rectifier output. It cannot charge up completely during the next output pulse. Between them both (the inductor and capacitor), they provide a good filtering action.

As a matter of safety, a bleeder resistor is usually connected across the condenser terminals to discharge the capacitor energy when the power supply is turned off.

A complete, simple arc welding power supply circuit is shown in Figure VIII-15. It employs a tapped transformer for controlling the voltage output and an inductor-capacitor filter.

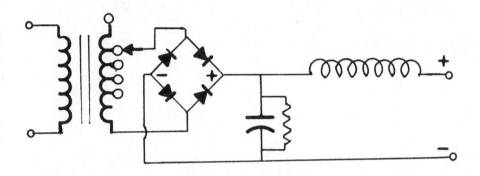

FIGURE VIII-15: Simple 1Ø FWB Rectifier Power Supply With Voltage Adjustment & Filtered Output

### Three Phase Rectification (3Ø)
With single phase rectifier circuits, it is necessary to use filters to reduce the ripple and smooth the output wave shape. Three phase rectifier circuits inherently have less ripple and do not require much, if any, filtering. Three phase rectifiers work in the same manner as single phase units. However, a detailed analysis of their operation is beyond the scope of this text. Their operation will be explained by way of a simple model, based on single phase principles, and an IMAGINARY conductor.

### Half Wave Rectification, 3Ø
As in single phase circuits, a half wave three phase rectifier circuit uses only 1/2 of the input waves. it blocks the other half waves. In three phase rectifier circuits, the rectified output is carried to the load by a single circuit loop which is connected between the plus and minus terminals of the rectifier.

*Although it is not possible, imagine* that all three phases of AC *could* flow in a single wire. Then we could take three sine waves and overlap them as shown in Figure VIII-16.

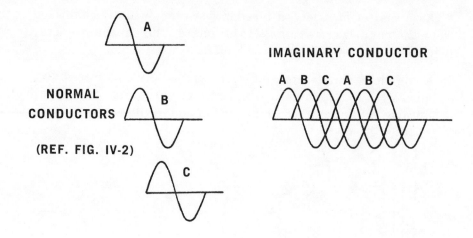

NORMAL
CONDUCTORS

(REF. FIG. IV-2)

IMAGINARY CONDUCTOR

A B C A B C

FIGURE VIII-16: Three Phase Currents

When a diode is placed in series with the imaginary three phase conductor, half wave rectification takes place. As in the half wave single phase case, shown in Fig. VIII-2, the bottom half of each sine wave is blocked. This is shown in Figure VIII-17, which is a repeat of Figure VIII-16 with the bottom half of the three sine waves missing.

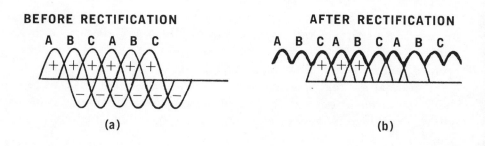

BEFORE RECTIFICATION

A B C A B C

(a)

AFTER RECTIFICATION

A B C A B C A B C

(b)

FIGURE VIII-17: Typical Output Wave Shape Of Three Phase Half Wave Rectifier

The heavy line represents the DC output from a half wave three phase rectifier circuit. Note that the ripple is small. The unfiltered half wave three phase rectified current does not go to zero, at any time, as it does in single phase systems. Figure VIII-18 compares the outputs of the single phase and half wave three phase systems. (See Figures VIII-2,-7,-17)

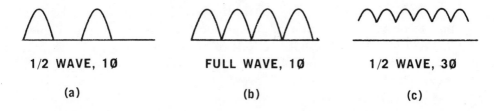

**1/2 WAVE, 1Ø**

(a)

**FULL WAVE, 1Ø**

(b)

**1/2 WAVE, 3Ø**

(c)

FIGURE VIII-18: Comparison Of Unfiltered Rectifier Outputs

An actual circuit diagram for a three phase 1/2 wave rectifier is shown in Figure VIII-19. Three diodes are used in the circuit. It shows one diode in *each* lead from a three phase Wye transformer. The neutral of the transformer is used as the negative terminal of the rectifier system.

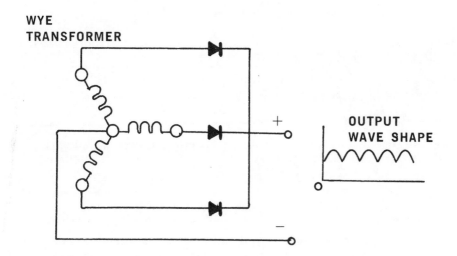

FIGURE VIII-19: Simple Three Phase Half Wave Rectifier Circuit

Each diode of the 1/2 wave 3 Ø circuit in Figure VIII-19 contributes one of the "bumps" in the output wave shape shown in Figure VIII-17 (b) and 18 (c).

### Three Phase Full Wave Bridge (3Ø FWB)

The amount of ripple in the output of a rectifier is reduced even further with full wave rectification of three phase power. Instead of "throwing away" the bottom half of the sine waves (as is the case with 1/2 wave 3 Ø rectifiers) full wave 3 Ø rectifiers use both halves of the sine wave.

Just as was done for the half wave case, *imagine* all three phases of AC flowing in a single conductor. This is shown in Figure VIII-20 (a). Take the bottom half of each sine wave (A,B, and C) and imagine them to be folded upward so they overlap the top half waves. This is shown in Figure VIII-20 (b).

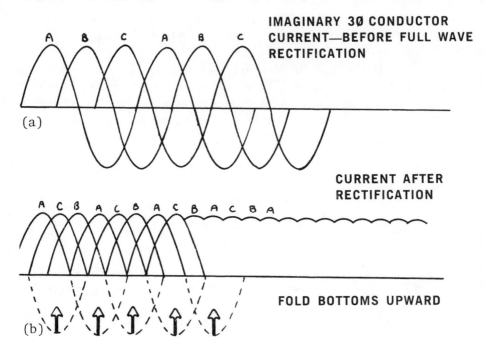

(a) IMAGINARY 3Ø CONDUCTOR CURRENT—BEFORE FULL WAVE RECTIFICATION

CURRENT AFTER RECTIFICATION

FOLD BOTTOMS UPWARD

(b)

FIGURE VIII-20

Instead of three peaks and valleys in the space of one sine wave (as is the case with 1/2 wave 3Ø rectification), there are 6 peaks and valleys. The ripple is much less with 6 overlapping half waves than with three half waves.

The rectified output wave shape for any number of AC phases can always be found by imagining that all the phases are flowing at one time, in one imaginary conductor; and either

1. Throwing the bottom half away for 1/2 wave rectification or,

2. Folding up the bottom half of each wave to overlap the top half for full wave rectification.

The output ripple decreases as the number of phases is increased and when full wave rectification is used instead of half wave. (See Appendix Table IV).

98

Figure VIII-21 shows a typical 3∅ Full Wave rectifier circuit. It is a basic circuit used in many three phase rectifier-transformer welding power supplies.

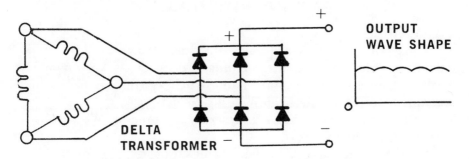

FIGURE VIII-21: Simple Three Phase Full Wave Rectifier Circuit

Note in Figure VIII-21 that there are 6 diodes. Full wave 3∅ rectifiers use twice as many diodes as are used in the 1/2 wave, 3∅ rectifier of Figure VIII-19. In this case, it doesn't matter if the transformer is connected in delta or Wye, the six diodes will rectify any 3∅, three wire AC input.

Figure VIII-22 shows the full wave wye connection version of Figure VIII-21. There are other versions of three phase rectifiers, but the 3∅ FWB rectifier is used by most manufactureres of welding power supplies.

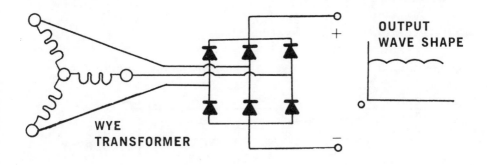

FIGURE VIII-22: Simple Three Phase Full Wave Rectifier Circuit

# CHAPTER IX

# TIG WELDING
# POWER SUPPLIES

## Introduction

Tungsten Inert Gas welding, otherwise known as TIG, is a nonconsumable electrode arc welding process. It is known by a number of different proprietary names such as Heliarc, Heliweld, Argonarc, etc. The American Welding Society uses the term *Gas Tungsten Arc welding* (GTA) to describe the process. We will use the term TIG in this text.

TIG welding uses nonconsumable tungsten welding electrodes. Tungsten is used because it has a melting point of 3370°C. It is the highest melting point of all the metals. The tungsten electrode is surrounded by an inert gas shield (argon, helium, etc.) to protect it from oxygen in the air. Otherwise, it would be consumed by the arc. When the current is great enough, the tungsten electrode melts at the arc end and forms a ball. Should the current be too great, the molten ball of tungsten at the end of the electrode will fall into the weld pool. Filler metal is added by means of straight rods. The TIG arc is used to melt metal from the end of the rod, as it is carefully guided into the welding zone. This chapter discusses the various power supply requirements for producing and controlling TIG arcs.

## The Elementary TIG System

As shown in Figure IX-1, an elementary TIG welding system consists of a power supply, a torch and some shielding gas.

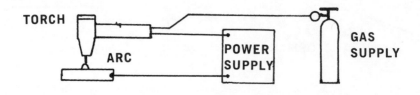

**FIGURE IX-1: Elementary TIG Welding System**

The welding a TIG arc can do depends a great deal on the electrical power delivered to the arc. The electrical watts that are converted to heat by the arc are used for welding. Once the shielding gas has been selected, it is the arc current and the arc length (voltage) which control the "heat" of the arc. The skillful selection of current and manipulation of arc length (voltage) are the means by which an operator controls the arc "heat". Automatic TIG welding systems also use controlled arc current and arc length to manipulate arc power.

## TIG Watts

The electrical power (watts) converted to heat by a TIG arc depends on the arc current and arc voltage. The current is determined by the power source setting and the arc voltage is essentially determined by the arc length. This can be demonstrated by way of the following illustration.

Imagine a TIG arc of approximately 100 amperes, operating in an argon gas atmosphere, and that the arc length causes the arc to operate at 15 volts. The welding condition can be represented as shown by the "X" in Figure IX-2.

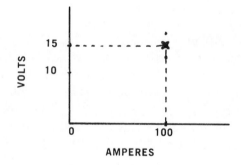

FIGURE IX-2: Example of TIG With Constant Current Power

In this example, the arc zone consumes 1500 watts* of electrical power. Some of this power is converted into useful heat which melts the workpiece, some into radiation and the rest of the power is lost (does no useful work).

An increase in the original 15 volt arc length increases the arc voltage. It takes more voltage to push the current across the larger arc gap. Assume the current remains constant at 100 amperes and that the arc length is increased to 17 volts. The arc power increases to 1700**watts. The arc becomes "hotter". Again, asume a constant 100 amperes. Now decrease the arc length to 13 volts. The arc power is now 1300 watts.

Table IX-1 summarizes the power changes which occur as the arc length is varied.

*100x15 = 1500
**100x17 = 1700

# TABLE IX-1

| Approximate Length Change | Voltage | Current | Power | Power Change |
|---|---|---|---|---|
| -0- | 15 volts | 100 amps | 1500 watts | -0- |
| +1/16″ | 17 volts | 100 amps | 1700 watts | +200 watts |
| −1/16″ | 13 volts | 100 amps | 1300 watts | −200 watts |

As can be seen in Table IX-1, a change in length of approximately ±1/16″ causes a change of power of approximately ±200 watts. The change in watts which accompanies a change in arc length, is one means by which an operator controls the weld pool.

A truly constant current power supply would provide the current required in the example.

### TIG Watts and Slope

Arc current will vary with changes in arc length, unless a constant current power supply is used. The preceding topic describes a constant current TIG arc and its response to changes in length (volts). This topic discusses the effect of power supply slope. Arc current varies depending on the amount of slope. The effect of slope can be demonstrated by way of another illustration.

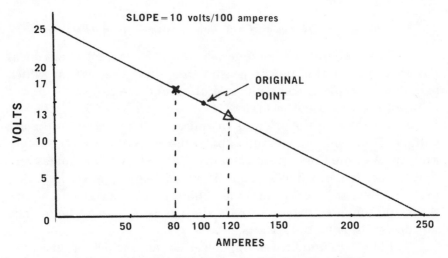

FIGURE IX-3: Effect Of Slope On TIG Operating Point

Imagine the same 100 ampere, 15 volt, TIG arc operating with power supplied by a power supply that has a 10 volt/100 ampere slope. The operating point for this new condition is the same one as shown by the "X" in Figure IX-2. As in the previous illustration, the arc zone consumes 1500 watts.

However, in this case, there is a difference. When the arc length is increased to 17 volts, instead of remaining at 100 amperes, the current changes. Because there is slope to the volt ampere curve of the power supply, the change in length causes a 20 ampere decrease in the arc current. This is shown in Figure IX-3, where the "X" is at 17 volts and 80 amperes (2 volts higher and 20 amperes lower than the original point).

The arc zone power decreases to 1360 watts,* instead of remaining at 1500 watts.

As in the earlier illustration, decrease the arc length to 13 volts. The current increases at the same time by 20 amperes, to 120 amperes. The arc zone power increases to 1560 watts.** This new operation point is shown by a Δ in Figure IX-3.

The change in length in the previous illustration and shown in Table IX-1 did not change the arc current. In this illustration, the current did change, from a low of 80 to a high of 120 amperes. Table IX-2 summarizes the power changes of this illustration when the power supply has a 10 volt/100 ampere slope.

### TABLE IX-2

| Approximate Length Change | Voltage | Current | Power | Power Change |
|---|---|---|---|---|
| -0- | 15 volts | 100 amps | 1500 watts | -0- |
| +1/16″ | 17 volts | 80 amps | 1360 watts | -140 watts |
| −1/16″ | 13 volts | 120 amps | 1560 watts | +60 watts |

As can be seen from Table IX-2, a change in length of approximately ±1/16″ causes a change in arc power of approximately −140 watts to +60 watts. A comparison of the examples in Tables IX-1 and IX-2 make it apparent that changes in slope cause changes in the power response of TIG arcs. A closer inspection, of the constant current example in Table IX-1, and the slope example in Table IX-2, will also show that the *direction* of

* 17 volts x 80 amperes = 1360 watts
** 13 volts x 120 amperes = 1560 watts

103

the power change has *reversed*. Instead of increased power with increased length, as in the constant current case, the arc power decreased with a length increase in the slope case.

The previous two examples were discussed to point out that a TIG arc will behave differently when the slope of the power supply is changed. It is even possible to have a volt/ampere curve (slope) which produces *no* change in power when the arc length is varied. With such power supplies, as the voltage is changed, the current is varied at the same percentage, in the opposite direction, keeping power constant.

Slope is an effective tool by which an operator can change TIG arc power response to changes in length. However, there is a point of caution. Note, in the example shown in Figure IX-3, that the short circuit current available from the power supply is 250 amperes. As the slope is made even more "flat", the short circuit current increases to above 250 amperes. Too much short circuit current is a problem when TIG arcs are started by the "scratch start" method. High short circuit current surge damages the tungsten electrode. A short circuit current of from 1.5 to 2 times the operating current is generally considered as safe.

Satisfactory power response and good "scratch" starting can be obtained with the conventional or CC type power supply. (See Chapter VII) The flat slope, or CP power supply, is not normally used for TIG welding because the short circuit current is too high.

### "Touch" Starting

Starting TIG arcs is not as simple as it may seem. The most obvious way to start the arc is to touch the electrode to the workpiece (to begin current flow) then raise the electrode from the workpiece. As the electrode lifts from the workpiece the current is kept flowing by inductance in the circuit. The small spark created between the electrode and workpiece, when contact is broken, grows into an arc. This method of starting a TIG arc is called "touch" or "scratch" starting. It has several drawbacks as shown in Table IX-3.

### TABLE IX-3
### TOUCH STARTING TIG ARCS

| Advantages | Disadvantages |
|---|---|
| 1. Simple to use | 1. Tungsten contamination of the workpiece. |
| 2. Non-mechanical | 2. Erosion of the Tungsten electrode. |

3. High initial arc current due to the short circuit.
4. Random time of start.
5. Not 100% sure of starting at each "touch".
6. Impractical to set arc length in advance.
7. Difficult to start arcs remotely.

Touch starting is an effective tool where simplicity is desired and a high quality weld is not essential. Other more complex methods are used when quality welding is desired.

### High Frequency Starting

A method of arc starting, which overcomes many of the "touch start" disadvantages, is High Frequency Starting. It is usually called HF starting. In this method, a high voltage (several thousand volts) with a high frequency (up to millions of cycles per second), is connected across the arc gap. The high voltage causes the arc gap to become ionized and break down. The size of the gap which can be jumped depends on how much voltage is used across the gap, as well as the type of gas, and the shape and material of the electrode. Properly adjusted HF equipment can readily start arcs of 1/8″ to 1/4″, and larger.

The high voltage would be a safety hazard if it were not for the high frequency. It is an inherent characteristic for HF current to flow on the *surface* of a conductor instead of through the cross section. Should a welder come in contact with HF, the energy will

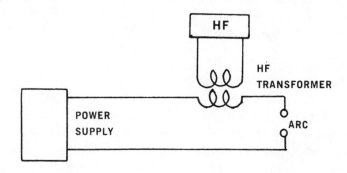

FIGURE IX-4: Schematic of HF Starting Circuit

travel over his skin instead of through it. He will feel a slight shock, but it will not cause a healthy person any trouble, except some discomfort.

There are many ways to put HF into a welding circuit. One way is to use a HF transformer as shown in Figure IX-4.

The HF energy is super-imposed on the welding circuit, through the HF transformer. The HF transformer winding in the arc circuit is made of several turns of large diameter wire. Its presence does not change the welding circuit, except for providing good arc starting. The advantages of HF are shown in Table IX-4.

## TABLE IX-4
### HF ARC STARTING

| Advantages | Disadvantages |
|---|---|
| 1. Reliable | 1. Can cause radio interference |
| 2. No electrode or workpiece contamination | 2. Requires some maintenance |
| 3. Safe | 3. Increased cost |
| 4. No starting short circuit current surge | 4. Circuit requires superior insulation |
| 5. Can be used remotely | |
| 6. Relatively simple | |

Arc starting may be influenced by some of the following conditions. For example, sharp pointed electrodes start easier than those with blunt ends. Dirty electrodes (touched by fingers, oxide contamination, etc.) are easier to start than clean electrodes. In fact, with very clean electrodes and with continuously flowing high purity inert shielding gas, after the first few arc starts, it becomes very difficult to start an arc.

**Other Starting Methods**

There are many variations of HF arc starting. All of the methods employ high voltage and high frequency or its equivalent. Some starting devices put out a single voltage spike, some use a train of pulses. Most of the variations are intended to overcome the radio interference due to HF. There is even a method which uses a pilot arc (a small, additional arc), much in the same way as a gas stove uses a pilot lite flame.

## Tungsten Electrodes

The current carrying capacity of pure tungsten electrodes can be increased by using some other material in the electrode. It is also possible to improve arc starting by using alloys of tungsten. There are a number of different tungsten (alloy) electrodes available commercially. Each of the alloys is claimed to have certain advantages over pure tungsten. However, pure tungsten is still a valuable welding tool. For example, pure tungsten will melt and form a ball on the end of the electrode. This improves arc stability in some welding applications. The main characteristics of two different Tungsten Electrodes are shown in Table IX-5.

### TABLE IX-5
### TUNGSTEN ELECTRODES MAIN CHARACTERISTICS

| Pure Tungsten | Thoriated Tungsten |
|---|---|
| 1. Lower current capacity | 1. Higher current capacity |
| 2. Forms a ball on the electrode end | 2. Maintains its shape, does not ball readily |
| 3. Less expensive | 3. Greater resistance to contamination |
| 4. Harder starting | 4. Easier starting |

## AC or DC?

Both AC and DC can be used to operate TIG arcs. DCRP heats the electrode end much more than DCSP. The extra heating effect is due to the direction of flow of the electrons.

The electrode is positive in the RP case. It is constantly bombarded by electrons and is heated because of the energy released by the electron collisions.

In the SP case, the electrode is negative and is constantly emitting electrons. In a sense, the electrons carry heat away from the electrode and keep it cool. The current carrying capacity of a Tungsten electrode is considerably higher with DCSP than with DCRP.

Tungsten electrodes with alternating current operate at an in-between heat level. Not quite as hot as RP, or as cold as SP. Because AC arcs operate as RP for part of the time, and SP for the remainder, there is an unbalance in the alternate half waves of current. Part of the time, the electrons flow from a point (electrode) to a plane (workpiece) and part of the time they flow in the opposite direction.

It is easier to push current through the arc in the SP direction than in the RP direction. The electrons can jump from the electrode tip more readily than from the workpiece. The same voltage applied to an SP arc will push more current than with an RP arc. When an AC voltage is used, the current becomes "unbalanced". This is shown in Figure IX-5. The current unbalance must be allowed for in the design of a TIG power supply. The unbalance shows up as a DC component of current in the transformer. Unless extra iron is provided in the core of the transformer, the DC will distort the AC output current wave shape and cause arc instability.

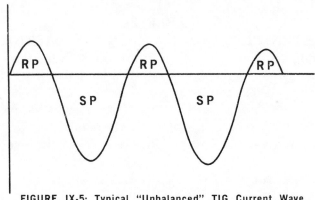

FIGURE IX-5: Typical "Unbalanced" TIG Current Wave

A simple alternating current TIG arc power supply is shown in Figure IX-6.

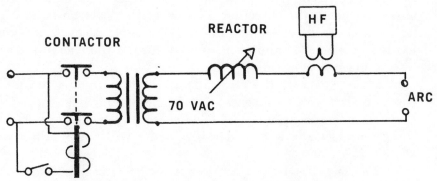

FIGURE IX-6: Schematic of a Simple TIG Power Supply

In Figure IX-6 when the single pole switch to the main contactor coil is closed, the main contactor energizes the transformer. The full 70 volts of the transformer is applied across the arc gap. The HF system breaks down the gap and starts the arc. The arc current is

108

controlled by the adjustable reactor. Inasmuch as an AC arc goes out at the end of each half cycle, the HF unit is not usually turned off. It is operated continuously, to reignite the arc each half cycle. At least twice the 70 VAC open circuit voltage (140 VAC) is usually needed before a TIG arc will *run* without the help of HF.

The addition of a full wave rectifier and filter circuit readily converts an AC power supply as shown in Figure IX-6, into a DC unit.

A very popular type of TIG power supply is the so called "AC/DC" unit. It can provide either AC or DC by changing a few switches or connections. Most AC/DC power supplies have the following features in common:

1. ON-OFF provision, by means of switches or contactors.
2. Current adjustment, by means of a variable reactor.
3. HF, for starting or continuous operation.
4. Switches, for changing current ranges or from AC to DC.
5. Auxiliary devices, such as gas and water solenoids, timers, meters, etc.

## Reactor Current Control

Each AC circuit in this chapter uses a variable reactor for current control. Adjustment of the reactor is used to change the amount of droop of the V/A curve and consequently the available arc current. Although resistors and condensers could be used to control the V/A curve, reactors are used because they produce a *Lagging* power factor.

Alternating current arcs work better with a lagging power factor, because of the phase shift between current and voltage. When the current passes thru zero, the arc extinguishes. The phase shift due to the reactor causes the voltage to be near its peak when the current is at zero, thereby helping re-ignite the arc. A resistive circuit (no phase shift) would have zero voltage at zero current, thereby making arc re-ignition virtually impossible. Circuits with leading power factor (capacitive circuits) cannot be used at all, because the ignition voltage is in the wrong direction.

Ideally, a 90 degree (purely inductive) phase shift is desired for best arc re-ignition. In practice, the phase shift of many power supply circuits approach 60 degrees. Things which add resistance, such as extra long cables, splices, undersized conductors, etc., decrease the phase shift from 90 degrees toward zero degrees and can cause poor arc re-ignition. It is good practice to minimize resistance in the welding circuit.

**A.C. Arc Rectification**

Alternating current TIG arcs go out each time the current passes through zero. Normally, the re-ignition process takes place so fast that the operator is not aware of the OFF instant. Sometimes an AC arc may fail to re-ignite after it has extinguished.

Usually, an AC arc will fail to re-ignite on the RP half cycle. The SP half is very easy to ignite and maintain. The RP half cycle is responsible for the "cleaning" action of AC arcs. When it is missing, especially on aluminum, the weld becomes dirty and the arc becomes virtually useless.

The term RECTIFICATION as applied to AC arcs refers to any of the things by which a DC component of arc current is produced. There are several types of AC arc rectification. These are summarized in Table IX-6.

TABLE IX-6

**TYPES OF AC ARC RECTIFICATION**

| | |
|---|---|
| 1. Inherent Rectification | The natural unbalance in an AC arc. See Figure IX-5. |
| 2. Complete Rectification | One arc half cycle is completely missing (Usually the RP half cycle). |
| 3. Partial Rectification | Late ignition of arc half cycles. |

Failure to re-ignite or late re-ignition can be due to many things, a few of which are listed in Table IX-7.

TABLE IX-7

**CAUSES OF RE-IGNITION FAILURE OF AC ARCS**

1. Arc gap becomes too long; not enough voltage to jump the gap.
2. Low voltage; the input to the power supply may be too low.
3. Current too low; not enough current to keep the electrodes hot.
4. Weak HF output; the HF unit needs servicing.

Complete and partial rectification of AC arcs can be kept to a minimum by careful practice and paying attention to details. Some guidelines are listed in Table IX-8.

## TABLE IX-8
### GUIDELINES TO MINIMIZE RECTIFICATION OF AC ARCS

| | |
|---|---|
| 1. Minimize circuit resistance. | 4. Maintain good arc shielding. |
| 2. Use the proper size electrode. | 5. Maintain equipment in good condition, especially the HF system. |
| 3. Keep the arc length short. | |

### Balanced Wave AC

The naturally occurring unbalance between the RP and SP half cycles of an AC arc can be controlled and even eliminated by *wave balancing*. The RP half cycle of current is increased and the SP half cycle is decreased. When the RP and SP half cycles are made to balance, the arc has a different heat distribution. The current carrying capacity of the tungsten electrode is also reduced. When the RP half cycle is increased, by wave balancing, the average heating effect of the arc on the electrode is also increased. Consequently, the electrodes have a lower current rating with balanced wave power*.

Balanced wave AC power is especially useful in the welding of aluminum and other metals that have a tenacious oxide layer. The RP half cycle of an AC arc "cleans" the oxide from the workpiece. As shown in Figure IX-7, balanced wave power has a greater RP component than an unbalanced wave. As a result, the "cleaning action" of the balanced wave arc is much greater. In addition to providing better cleaning action, balanced wave power supplies do not need the extra iron in the transformer. The DC component of current disappears when the wave is balanced and a normal amount of iron is satisfactory.

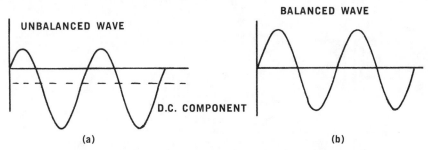

FIGURE IX-7: Comparison of "Unbalanced" and "Balanced" Waves

* See Appendix Table V for Tungsten Electrode Ratings.

A simple, yet effective way, of balancing the arc current is to place a large condenser in series with the arc.* This is shown in Figure IX-8.

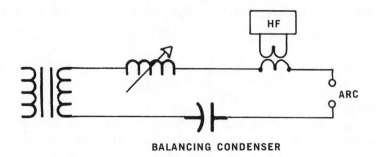

**BALANCING CONDENSER**

**FIGURE IX-8:** Typical Condenser Balanced Wave Power Supply

Direct current cannot pass through a condenser. The DC which *tries* to flow thru the circuit charges the condenser opposite in polarity to the normal SP voltage. The charge helps the current flow in the RP direction and opposes the flow in the SP direction. It opposes and aids current flow at exactly the correct amount, automatically, and it always produces a completely balanced current wave.

Another way to help balance an AC arc is to place a battery in series with the arc. The battery is connected to help current flow in the RP direction and oppose current flow in the SP direction. Unlike the "balancing condenser", a battery cannot change its amount of "balance" automatically. Therefore, with a battery balancer, the current tends *toward* a balanced condition but seldom achieves it. A slight unbalance usually remains. The battery balance system is shown in Figure IX-9.

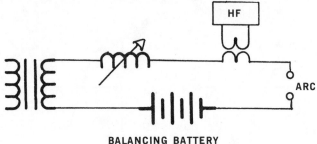

**BALANCING BATTERY**

**FIGURE IX-9:** Typical Battery Balanced Wave Power Supply

---

* The large condenser does not change the phase shift from *lag* to *lead*. The effect of the current adjusting reactor still dominates the circuit and produces a *lagging* phase shift.

The battery system needs some auxiliary equipment to take care of the battery, such as a charging system, etc. The condenser type of balance system needs no extra care.

A third way to help balance arc current is to add resistance to the circuit as shown in Figure IX-10. It is not a very effective way of balancing the current, but it does produce some balance of the current.

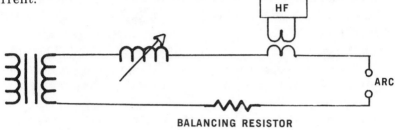

BALANCING RESISTOR

FIGURE IX-10: Partially Balanced Power Supply With Resistor

The resistor can also be used in conjunction with a half wave rectifier (diode), so that it is only in the circuit for half of the time.

Figure IX-11 shows the resistor is bypassed by a diode when current flows in the RP direction (clockwise) and is in the circuit when the current flows in the SP direction. The circuit reduces the SP current while the RP remains untouched. The tendency is to produce a more balanced wave. However, the amount of balance varies with the amount of current flow.

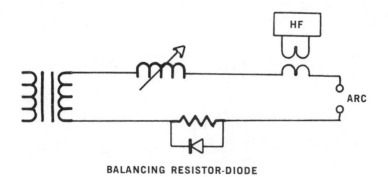

BALANCING RESISTOR-DIODE

FIGURE IX-11: Partially Balanced Power Supply With Resistor-Diode System

# CHAPTER X

# PLASMA ARC POWER SUPPLIES

## Introduction

Plasma arcs, like TIG arcs, are another form of non-consumable electrode arc systems. In a sense, a plasma arc is nothing more than a high power TIG arc, squeezed thru a small hole. Plasma arcs are quite a bit longer than TIG arcs. Consequently, the electrons (current) have further to travel. Therefore, plasma arcs run at higher voltage than TIG arcs and they are harder to start. In general, plasma arcs operate according to the same rules as TIG arcs.*

The arc power is concentrated by squeezing the arc plasma** through a small hole. The arc force is also increased by the "squeeze".

Just as TIG arcs have a wide range of operating conditions, so do plasma arcs. Control of the arc power and the degree of squeeze make plasma arcs useful for welding and surfacing, as well as cutting. This chapter discusses the power source requirements for producing and controlling plasma arcs.

## Transferred or Nontransferred?

There are two basic types of plasma arcs; these are 1) the transferred and 2) the nontransferred.

The *transferred* plasma arc is almost identical to a TIG arc. The plasma arc is struck and maintained between the electrode (in the torch) and the workpiece. The arc is squeezed through a small water cooled nozzle, attached to the torch as shown in Figure X-1.

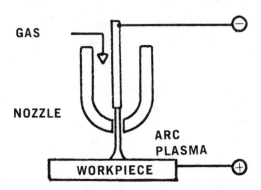

GAS

NOZZLE

ARC
PLASMA

WORKPIECE

FIGURE X-1: Main Elements of A Transferred Plasma Arc

* See Chapter IX
** This is the origin of the term "plasma arc".

The nozzle is electrically isolated from the arc. The current flows from the electrode to the work, thru the nozzle.

The *transferred mode* is particularly suited to metal cutting, as well as welding and surfacing.

The *nontransferred* arc is slightly different from the transferred arc. It is connected (electrically) between the torch electrode and the water cooled nozzle itself. The plasma is blown through the nozzle hole by the expanding hot shielding gas. It looks like, and acts like, a flame. However, it is much hotter. The current flows from the electrode, through the plasma, to the water cooled nozzle as shown in Figure X-2. There is no current flow to the workpiece. Nontransferred arcs operate at low power when compared to transferred arcs. A considerable amount of power is wasted as heat, which is lost to the water cooled nozzle. The *nontransferred mode* is suited for welding and surfacing.

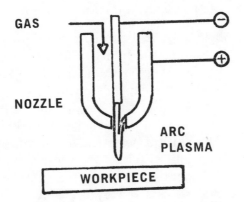

FIGURE X-2: Main Elements of a Nontransferred Plasma Arc

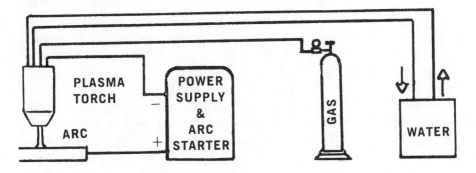

FIGURE X-3: An Elementary Transferred Plasma Arc System

### The Elementary Plasma Arc System

Most plasma arc systems are of the transferred type, as shown in Figure X-3. An elementary transferred plasma arc system consists of a power supply, a torch, a gas source, and a water source. Figure X-4 shows an elementary nontransferred plasma arc system.

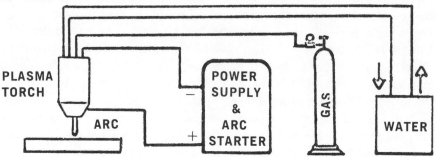

FIGURE X-4: An Elementary Nontransferred Plasma Arc System

The work a plasma arc can do depends on the amount of electrical power which is converted to heat. Plasma arc voltages are of the order of 2 to 10 times higher than those of TIG arcs. The actual voltage of a plasma arc depends on such things as the type of shielding gas, gas flow rate, the nozzle hole size and the arc current. The current range of a plasma arc is about the same as that of TIG arc. The skillful selection of plasma arc length, gas flow etc., in combination with the amount of arc current is the means by which an operator controls the plasma arc power.

Plasma arcs may be started by any one of several devices. These devices include High Frequency units, impulse systems, mechanical probes, pilot arc systems, etc. Considerable arc starting knowhow has been developed by the firms which manufacture plasma arc systems. However, once the arc is started, the power supply requirements are about the same for the various commercial plasma arc systems.

### Plasma Arc Power Sources

Most plasma arc power is direct current straight polarity. One reason for this is that DC plasma arcs are easier to maintain than AC plasma arcs. The most commonly available plasma arc power supply is the constant current (CC) or "drooper" type.* The major difference between a plasma arc unit and a TIG unit is the voltage output. The voltage of the plasma unit can be as high as 400 volts.

The high voltage can be obtained from especially designed

* See Chapter VII

116

power supplies or by the series connection of TIG units. In fact, there are some plasma units which may be reconnected internally to provide several different output curves.

Figure X-5 illustrates the V/A curve of a typical plasma power supply used for cutting applications. Its rated ouput is 200 volts at 250 amperes. In this case, the arc operates at approximately 10 times the voltage of typical 250A TIG arc. Plasma arc power can also be obtained by connecting several TIG power supplies together.

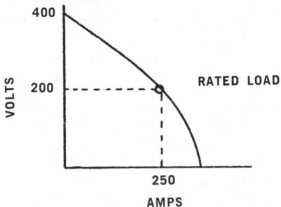

FIGURE X-5: V/A Curve of a Typical Plasma Arc Power Supply

Figure X-6 shows the composite V/A curve of several TIG power supplies. The TIG units are connected in series to provide an increased voltage.*

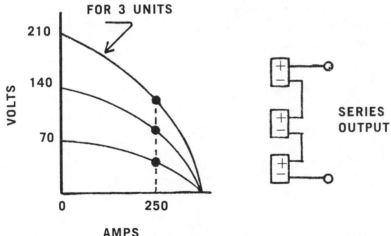

FIGURE X-6: V/A Curves of Three Series Connected Power Supplies

* Power sources should not be connected in series or parallel unless authorized by the manufacturer.

The voltage at each current on the curve for three units is 3 times the voltage of one unit. The current rating for three units connected in series is the same as that of a single unit.

Figure X-7 shows the V/A curve of three parallel connected units. At each voltage, the current is 3 times that of a single unit. The rated load voltage is the same as that of a single unit.

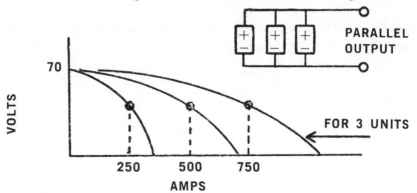

FIGURE X-7: V/A Curves of Three Parallel Connected Power Supplies

The combination of series and parallel connected power supplies will provide high voltage as well as high current. This principle is used by some manufacturers to build their plasma arc power supplies. Figure X-8 shows the family of V/A curves for a plasma unit with 4 separate, 100 volt open circuit, power circuits. The circuits may be connected all in series, parallel, or two by two.

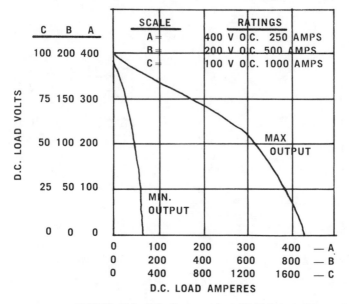

FIGURE X-8: V/A Curves of a Multi-Rated Power Supply

118

When the sections are series connected, the unit has a 250 ampere, 200 volt rating under load. When connected in parallel, the sections are rated at 1000 amperes and 50 volts. The "two x two" connection provides 500 amperes at 100 volts load. The open circuit voltages are 400, 100, and 200 respectively.

## Kilowatts or Amperes?

Some plasma arc power supplies are rated in KW instead of in AMPERES. Instead of saying the load condition is 200 volts and 250 amperes, or 75 volts and 100 amperes, etc. etc., the load is said to be 50 KW, or 7.5 KW respectively. There are some plasma arc power supplies which have been calibrated in terms of KW at rated voltage, instead of current at rated voltage.

There is no difference between a KW rated unit and an AMPERE rated unit. It is only a matter of how the indicator scale (dial) on the power supply is labeled.

# CHAPTER XI

# COVERED ELECTRODE WELDING
# POWER SUPPLIES

## Introduction:

Covered Electrode Welding, otherwise known as Stick Electrode Welding, is a consumable electrode welding process. It is also known as the Shielded Metal Arc Welding process by the American Welding Society. We will use the term *Covered Electrode* welding in this text.

The coatings on covered electrodes were developed to do two main things. One was to protect the molten metal from contamination by air, and the other was to stabilize the arc to make it easier to "keep alive". A bare steel rod cannot be easily used as an electrode when making a weld in air. The molten metal becomes contaminated with oxygen and nitrogen from the air. Not only that, the arc is almost impossible to sustain.

Covered electrodes are available in an almost endless variety for welding all types of metals. They are easy to use and do not require much skill to keep the arc going. The coatings have been developed to such a degree that almost any kind of constant current power source can be used for covered electrode welding. Alternating current as well as direct current can be used.

This chapter discusses the general features of power sources for covered electrode welding.

## The Covered Electrode System

The typical covered electrode welding system, as shown in Figure XI-1, consists of a constant current power supply, and an electrode holder. The covered electrode is held by the jaws of the holder, which is electrically connected to the power supply by a cable. The electrode holder is sometimes called a "stinger". An arc is started by "scratching" the electrode against the workpiece.

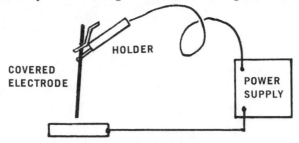

FIGURE XI-1: Typical Covered Electrode Welding System

The arc power is controlled by changing the current and arc length, in the same fashion as in a TIG arc (See Chapter IX). Current is determined by the setting of the CC power supply. Voltage is determined by the arc length and composition of the electrode covering. Once the current is set at the power supply, the operator need only feed the electrode slowly toward the weld puddle in order to maintain the required arc length. Too little feed of the electrode will cause the arc to lengthen until it goes out. Too much feed will make the electrode drag on the weld pool and cause a poor weld, and a great deal of spatter.

## Power Supply Types

In general, the same types of CC power supplies that are satisfactory for TIG are fine for covered electrodes. In fact, most TIG power supplies are sold as being satisfactory for covered electrode welding. The reverse is not necessarily true because TIG arcs are a little more critical in regards to current control. TIG welding usually requires a vernier adjustment of current, whereas the current may be adjusted in large increments with covered electrodes. The increments are selected to go with the standard sizes of covered electrodes.

The power supplies may be AC or DC, depending on the type of electrode covering. However, they are all of the constant current, or conventional type. Constant Potential power is not used for covered electrode welding.

## Spatter and Metal Transfer

The arc melts the end of the covered electrode and forms a molten drop of slag and metal. The drop is usually bounced around by the force of the arc, as it moves violently around the end of the electrode. The covered electrode arc is not quiet and steady as a TIG arc; it moves around constantly.

Sooner or later the bouncing molten metal drop, as well as melted covering (flux), transfers to the weld pool. At times, surface tension can cause the drop to bridge the gap between the electrode and the weld pool. When this happens, the arc goes out. Without a gap there can be no arc. The molten drop acts like a short circuit on the power supply. The voltage which maintained the arc, now pushes more current thru the circuit, unless the power supply is a truly constant current machine. The extra current heats the drop (metal bridge) even further, and causes it to "blow" just as a fuse blows when it is overloaded. Metal is scattered all over as a result of the fuse action. The fuse action is a main source of weld spatter. A similar thing happens with MIG welding, and will be discussed in detail in the chapter on MIG welding. Figure XI-2 shows a typical "fuse action" and the resulting spatter.

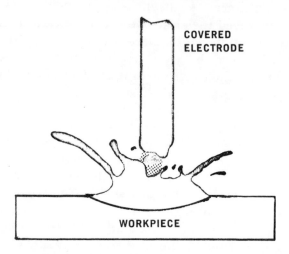

FIGURE XI-2: Typical "Fuse action" Spatter

The amount of spatter is controlled by the violence of the fuse action as well as the surface tension forces. The surface tension forces can be controlled to some degree by chemicals in the electrode coating. Some coverings cause less spatter than others. The fuse action is primarily controlled by the power supply. A rapid fuse action causes more spatter than a slow one. The fuse action is slowed down by adding an inductor to the power supply. A good DC covered electrode power supply has a large inductor in it to control the violence of the fuse action, and improve arc stability.

### Arc Dig Circuits

When the V/A curve of a covered electrode power supply is very steep, or even vertical, there is no appreciable change in current between the arc level and short circuit level. Consequently, the "fuse action" may be too weak to break the molten metal bridge between the electrode and weld pool. In that case, the electrode stubs into the pool and the pool freezes. Once electrode "stubbing" happens, the weld is ruined. Too little "fuse action" is prevented by extra circuits called "drive" or "dig" circuits. These circuits deliver extra current to the short circuit. They help give a good fuse action and eliminate electrode stubbing. The arc drive

122

circuits are nothing more than low voltage, high current, low duty cycle power supplies connected in parallel with the main supply. A typical circuit is shown in Figure XI-3.

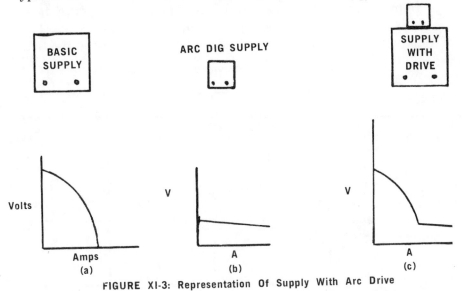

FIGURE XI-3: Representation Of Supply With Arc Drive

It is possible for V/A curves as shown in Figure XI-3 (c) to be designed right into the main power supply. No matter what the design, the reason for the extra current at low voltages is to get a good fuse action to prevent freezing of the electrode in the weld pool. Power supplies which have short circuit current between 1 1/2 to 2 times the arc current do not usually need arc dig circuits.

### The MO (Multiple Operator) System

The typical covered electrode arc operating time in any ten minute period is quite small. It is of the order of 20 to 30 percent. In locations with a large number of welding setups, a low equipment (arc) duty cycle wastes money and space.

Multiple Operator equipment helps solve the low utilization problem by using one machine to supply several operators. The MO technique of using one machine to supply several welding arcs, simultaneously, is satisfactory for most covered electrode welding. It is not satisfactory for MIG welding. The arc dynamics of MIG welding are too demanding.

A typical MO system is shown in Fig XI-4. A single constant potential power source of about 70-80 volts open circuit is used to supply the current. Each operator's station has a resistance box, with adjustable settings, for controlling the current.* As an example, assume a 25% operating time for each arc, and that there are 10 arcs operating at 200 amperes. Instead of using 10 individual 200 ampere power supplies, for a total of 2000 amperes, it is possible to use one 500 ampere power supply for all the arcs. The 500 amperes comes from 25% of the 2000 amperes of individual capacity.

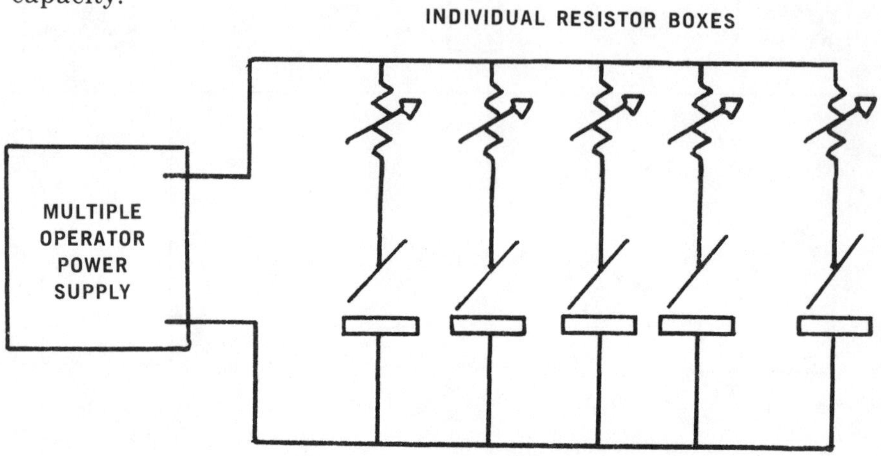

INDIVIDUAL RESISTOR BOXES

FIGURE XI-4: Typical MO Covered Electrode Welding System

The main disadvantage of the MO system is the loss of power in the resistors used to control the arc current. The efficiency of MO systems is not as good as single operator systems which use reactors for current control.

* See Example II-3, Chapter II.

CHAPTER XII

# MIG WELDING POWER SUPPLIES

## Introduction

MIG welding, like covered electrode welding, is a consumable electrode welding system. MIG is an abbreviation for *M*etal *I*nert *G*as welding. It is referred to as GMA (*G*as *M*etal *A*rc Welding) by the American Welding Society. The *Inert Gas* part of the MIG name is not technically correct. MIG welding uses *reactive* as well as *inert* shielding gases. Although early work with MIG welding used inert gases, this is no longer true. Reactive gas mixtures of argon & oxygen, as well as carbon dioxide and its mixtures, are widely used today. However, MIG remains the most popular name used to describe this consumable electrode welding system. This chapter discusses the general principles of MIG welding.

A MIG welding system is relatively simple. It requires a d. c. power supply, a welding torch with a contact tube, a spool of wire, a wire feeder and some shielding gas. The shielding gas is used to protect the molten weld metal from atmospheric contamination, just as in TIG welding. Power is transferred to the arc through the wire electrode as it is fed through the contact tube in the torch. With this welding process, the electrode is melted by the arc. The melted metal is transferred thru the arc to the work being welded. The wire feeder is used to feed more electrode toward the arc, just as fast as the arc melts the wire; otherwise, the arc would lengthen until it goes out. The average arc length remains relatively constant because more electrode is fed into the arc while it is melted by the arc.

In general, MIG arcs operate over the range of approximately $25\pm10$ volts d.c. and with about the same currents as covered electrode arcs.* The heat of a MIG arc is controlled by the proper selection of materials (wire and gas), the arc length and current, as well as type of power supply. The wide selection of gases, electrodes, and power supplies, are combined to form the various commercial welding systems. A typical welding system is illustrated in Fig. XII-1.

Each MIG system has its own merits and special area of application, but they all operate according to the same fundamental principles. These principles are discussed in the following pages.

* 60 to 600 amperes

125

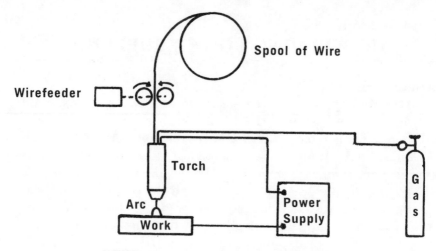

Spool of Wire

Wirefeeder

Torch

Arc

Work

Power Supply

Gas

FIGURE XII-1: Typical MIG Welding System

Special emphasis is placed on the role of the power supply in the effective use of MIG welding systems.

### Melting the Electrode

MIG welding depends on what happens to the molten metal drop on the end of the electrode. Virtually everything concerning MIG welding depends on that molten drop. This can be illustrated by way of the following example.

Imagine an electrode with its end partially melted by an arc and that a drop is beginning to form as shown in Figure XII-2.

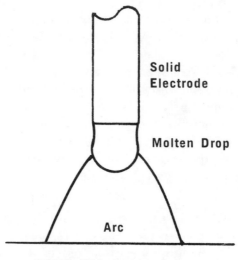

Solid Electrode

Molten Drop

Arc

FIGURE XII-2: Forming A Drop

In order to protect the molten drop from the air, it must be covered by a shielding gas. In the case of steel, the gas is usually argon or mixtures of argon with carbon dioxide or oxygen. Other gas combinations are sometimes found useful. Aluminum drops require inert gases for shielding, such as argon and helium. The type of gas which is actually used is of no great concern in the following discussion.

There are *two* generalizations which can be made from an examination of Fig. XII-2. *First*, to increase the number of drops formed in order to get more metal transfer, more arc current and wire is needed. An increase of the arc current causes the electrode to melt faster. Assume for a moment that all of the molten metal is transferred from the end of the electrode as fast as it is created. Unless more solid electrode is moved down, toward the workpiece, the arc length will increase and eventually cause the arc to go out. The welding system shown in Figure XII-1 is designed to feed the electrode at just the right speed to maintain a constant arc length. The arc length is constant when the feed rate and melt rate are equal. When the feed rate or melt rate is erratic, the arc length changes erratically. The dependency of feedrate (or meltrate) on current is usually shown by a graph of feed rate vs. current, as is illustrated in Figure XII-3. Feed rate is the same as melt rate when the arc length is steady.

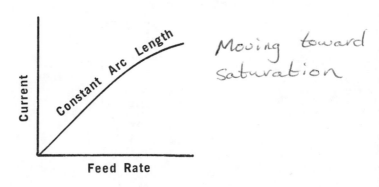

FIGURE XII-3: Typical Wire Burnoff Characteristic

*Second*, imagine a change in the electrode diameter. Larger electrodes must be fed slower than smaller electrodes, to get the same arc current. It is a simple matter of geometry. Figure XII-4 illustrates this concept.

127

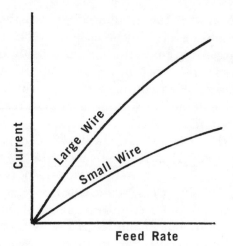

FIGURE XII-4: Effect Of Wire Size On The Wire Melt Rate

The two generalizations just discussed, and represented by Figures XII-3 and -4, apply to all consumable electrode processes. The differences between the processes are, in a large degree, determined by *how* the drop of molten metal is removed from the electrode end and transferred to the weld puddle.

## Metal Transfer

Basically, there are *three* ways for the drop to leave the end of the electrode. One way is for it to grow large enough in size that it falls off due to its own weight. This is called *globular transfer* and is illustrated in Figure XII-5.

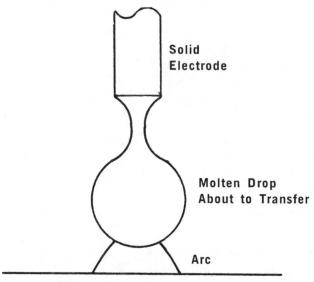

FIGURE XII-5: Example Of Globular Transfer

Globular transfer is limited in its use because it cannot be used for welding out of position. The heavy drop will fall down instead of transferring across to the workpiece, as is shown in Figure XII-6.

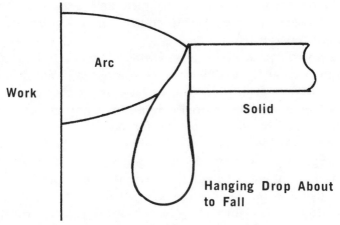

FIGURE XII-6: Example Of Globular Transfer In The Horizontal Position

A second way for the drop to be transferred to the weld puddle is by repetitive (intermittent) direct contact. This is called *short circuit metal transfer, or short arc* as it is sometimes known. Gravity plays a minor role in short arc transfer. Surface tension and "fuse action" are the biggest factors in this type of transfer. As shown in Figure XII-7, the electrode is made to touch the weld puddle because the melt rate fluctuates. When it is greater than the feed rate, the arc gets longer. When it is less than the feed rate, the drop is forced into the weld puddle.

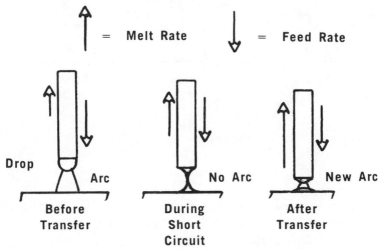

FIGURE XII-7: Example Of Short Circuit Transfer

The length of the melt rate and feed rate arrows illustrate *which one* is biggest and *when* it is biggest. The repetitive short circuit action cannot occur unless the fluctuating melt rate varies above and below the steady feed rate. Since melt rate changes with current, the power supply is used to control the melt rate by control of the arc current.

Because the drops are transferred when they touch the puddle, and not by gravity, it is possible to weld in all positions. It is easy to weld vertically, or even overhead, with short arc transfer. *Short arc transfer* is discussed in more detail in Chapter XIII.

The third way for drops to transfer is called *spray transfer.* With spray transfer, small drops, about the same size as the short arc drops, are transferred across the arc gap without short circuits. The key to spray transfer is the so-called "pinch effect". The pinch effect automatically squeezes the drops off the electrode before they get too big. Normally, the drops are pinched off when they are about the same diameter as the electrode. The pinch force is due to the *electromagnetic* effects of current. Spray transfer is illustrated in Figure XII-8.

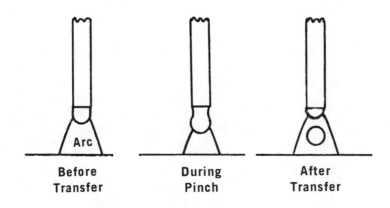

**Before Transfer**     **During Pinch**     **After Transfer**

FIGURE XII-8: Example Of Spray Transfer

There is a specific pinch current that must be reached before spray transfer occurs. The pinch current level is called the *transition current.* When the transition current is reached, the pinch force becomes greater than the surface tension force holding the

130

drop on the end of the electrode. As soon as the transition current is exceeded, the pinch force squeezes the drop off the electrode. It also gives the drop a strong *push* across the arc gap. The *push* given to the drop propels it across the arc gap and into the puddle. It makes it possible to weld in all positions, provided of course that the weld puddle is not so big that it won't stay in place. It is even possible to transfer metal in the overhead position. Other forces, due to gravity and gas flow, play a part in the drop transfer process, but their effect is usually too small to be noticed. The two biggest factors in spray transfer are *surface tension* and the *transition current*. Different sizes of wire, as well as types of wires, have different transition currents. Bigger diameter drops need more "pinch", therefore, the transition current goes up or down according to wire diameter. Some transition currents are shown in Table VI of the appendix.

## Surface Tension

Just as "surface tension" holds a drop of water on the end of a medicine dropper, it also holds a drop of molten metal on the end of an electrode. The strength of the surface tension force varies with the type of metal. Some metal drops "hold on" stronger than others. For example, aluminum has about half the surface tension of steel. An aluminum drop holds on only half as strong as a steel drop. The surface tension also depends on the type of gas which surrounds the drop. When argon (an inert gas) is used the surface tension is uniform and about the same as listed in reference books. The surface tension changes when oxygen (a reactive gas) touches the surface of the liquid metal. This is one reason why metal transfer becomes erratic when air gets into the arc zone. Change in surface tension is one of the things responsible for the "puckering", or "rosebudding" as it is sometimes called, of aluminum weld beads. As long as good gas shielding is maintained, and as long as the proper gas is used, the surface tension will be held constant. As long as the surface tension is constant, the metal transfer will be stable.

Experience has shown that argon, with a small amount of oxygen, gives a stable value of surface tension for both steel and aluminum. The *transition* current is easy to determine with these gases. When too much oxygen (oxides) is present the surface tension becomes erratic. Consequently, the transition current will also be erratic.

## Electrode Extension

The piece of electrode between its last point of electrical contact with the contact tube and the arc is the "electrode extension"; it is sometimes called the "stickout". This is shown in Figure XII-9.

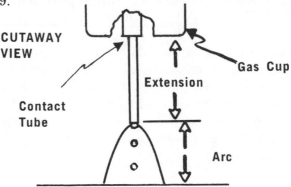

FIGURE XII-9: Definition Of Electrode Extension

The current which flows to the arc flows thru the extension. The extension is heated by the $I^2R$ energy developed. Because $I^2R$ energy heats up the electrode, it has an effect on the total melt rate. Increasing the stickout increases the melt rate because it also increases the extension resistance. Anything which changes the resistance of the extension changes the $I^2R$ energy, as well as changing the voltage required from the power supply. MIG arcs with long extensions need more voltage. Arcs with small diameter electrodes use more voltage than those with large electrodes at the same current.

In fact, as far as the arc is concerned, the resistance of the stickout is part of the *slope*\* of the power supply system. Most MIG arcs are started by running the electrode end into the workpiece. When the electrode touches the work, current begins to flow thru the extension, assuming, of course, that the power supply is energized. The $I^2R$ energy generated by the surge of starting current heats up the stickout, melts the electrode and pinches it off.

When the end of the electrode has a sharp point, or a jagged end, the arc can start by flashing off the point. The point acts as if it were a very small diameter electrode. It is heated by $I^2R$ energy and fuses off in the same way as is described in the preceding paragraph. A normal start is called a FUSE start. When a sharp point starts the arc, it is called a FLASH start. However, both types of starts use $I^2R$ energy to do the job.

\* See Chapter VII

## Pinch Effect

One of the most important factors controlling metal transfer is the "pinch" or "squeeze" on the electrode. Every wire that carries current is squeezed by the pinch force. If the pinch were great enough it could even change the shape of a solid wire. The strength of the pinch is related to the strength of the magnetic field set up around the wire, by the current flowing thru it. It is the same type of force that causes a welding cable to "jump" when a sudden surge of current flows.

For a given wire diameter the pinch (P) is proportional to the square of the current (I). This means that when the current is doubled the pinch goes up four times. When current is tripled, the pinch is nine times as strong. Figure XII-10 illustrates the concept of pinch effect. It shows a liquid metal bridge in the act of necking down.

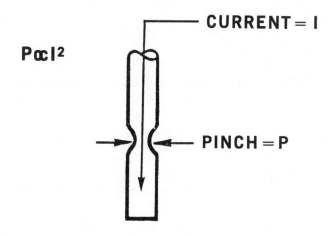

$$P \propto I^2$$

CURRENT = I

PINCH = P

FIGURE XII-10: The "Pinch" Concept

Spatter produced during short arc welding, including $CO_2$ welding, is a result of the pinching action. This can be explained by the following example. Assume there is plenty of pinch force and that the short circuit metal bridges, shown in Figure XII-7, are pinched off. The amount of spatter generated depends on the rate of pinch. The faster the pinch is applied, the more the spatter that occurs.

The concept of "rate of pinch" can be illustrated by use of an open toothpaste tube. When the tube is squeezed slowly, paste oozes out. When the tube is given a fast squeeze, paste squirts out. The faster the squeeze, the further the squirt. When the *amount* of squeeze is enough to *pinch* the bridge, it is the *rate* of *squeeze* which controls the amount of spatter. The faster the squeeze the more the spatter. This is illustrated in Figure XII-11.

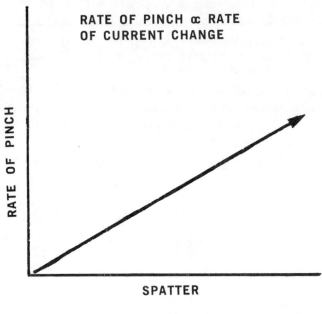

FIGURE XII-11

### Starting The Arc

As was pointed out in the section on *Electrode Extension*, $I^2R$ heating and pinch control the starting cycle. Both the heat and pinch depend on current. Therefore, in order to get a good arc start, it is necessary to control the current. First, the *amount of current* must be controlled, in order to control the *amount of pinch.* Second, the *rate of current* rise must be controlled in order to control the *rate of pinch.* Control over the current is obtained by use of the correct type of power supply. Figure XII-12 shows a typical arc start circuit.

The amount of current which flows is determined by the circuit resistance and output voltage of the power supply. When there is too much current, there is too much pinch, and the wire will literally explode; just as a fuse will vaporize when there is too much current.

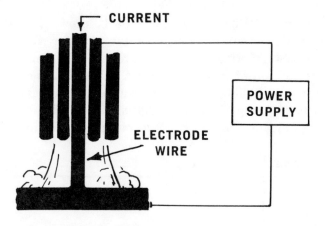

FIGURE XII-12: Typical "Fuse" Start

One way to keep spatter low is by limiting the current with power supply slope*, as is illustrated in Figure XII-13.

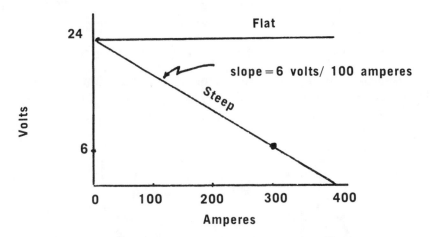

FIGURE XII-13: Slope Control Of Pinch Current During Starts

Assume that the power supply of Figure XII-12 has 24 volts open circuit, and a slope of 6 volts per 100 amperes. If the resistance of the electrode extension happens to be 0.02 ohms, then 6 volts will develop across the extension and 300 amperes will flow thru it. One way to prove this to yourself is to superimpose the load line of a 0.02 ohm resistor (see Chapter V, Figure 3) on the power supply

* See Chapter VII.

curve shown in Figure XII-13. This is shown in Figure XII-14.

If the power supply slope was flat, in other words, if there was no slope, the starting current would be 1200 amperes. The 1200 amperes are found from OHMS LAW by dividing the 24 volts by 0.02 ohms. An 0.030'' diameter steel electrode can be pinched by either the 300 amperes or the 1200 amperes, but a great deal more spatter would develop with a 1200 ampere starting current surge.

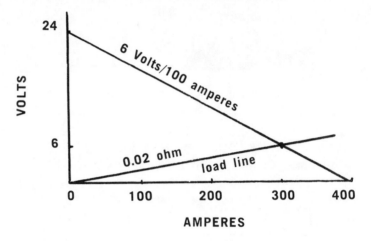

FIGURE XII-14: Load Line Method of Determining Starting Currents

Spatter can be minimized by using as much slope as possible. However, avoid too much slope, it will cause a "birds nest" or "ball of spaghetti" as shown in Figure XII-15.

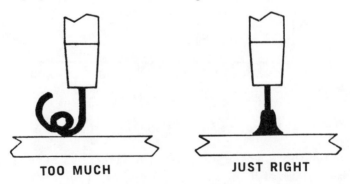

TOO MUCH          JUST RIGHT

FIGURE XII-15: Effect of Slope on Arc Starting

When slope is used to minimize spatter, there is one severe drawback to consider. It limits the voltage available from the power supply, depending on the amount of slope. Without slope, the full output voltage is always available. With slope added, the voltage is reduced. In cases where the reduction in available voltage is not important, use of slope for reduction of spatter is very effective.

The use of INDUCTANCE is another way to reduce spatter. It can control the rate of rise of the starting current surge.* When the correct amount of inductance is used, the *rate* of pinch is controlled, and spatter is minimized. Inductance does not limit the voltage available from a d.c. power supply. It only effects the rate of current rise, not the amount of current. Figure XII-16 illustrates the effect of inductance on the current rise. A combination of both slope and inductance will provide optimum starting characteristics.

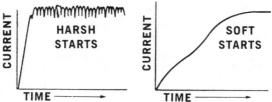

FIGURE XII-16: Typical Arc Start Current Vs. Time Oscillograms

In summary, the power supply requirements for MIG arc starts are as follows:

1) Controlled rate of pinch, by controlled rate of current buildup.
2) Controlled amount of pinch, by controlled amount of current.

The first requirement can be met by using INDUCTANCE in the weld circuit. The second can be met by use of SLOPE. Best starting occurs when both requirements are met at the same time.

## Separation and Control

In the previous section, *Starting the Arc,* it was pointed out that both the amount of current and rate of rise of current are important. These same effects are also important in running the arc. It is best to have separate control over both effects. In a sense, control of the amount of current is a *static* requirement, and control of the rate of current rise is a *dynamic* requirement. Separate control of the static and dynamic characteristics in a MIG power supply is of great value. When the system is set right, the arc starts are quieter and have less spatter, and the arc operation is also at its best.

* See Chapter II, Figure 16.

Figure XII-17 shows the block diagram of a rectifier-transformer type of power supply that has separate control over the slope (statics) and inductance (dynamics).

The slope and inductance are shown as variable because the correct settings depend on the type of material that is welded, the electrode diameter, the gas, etc.

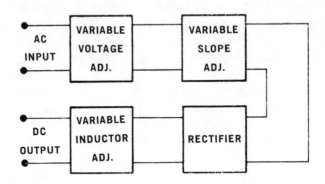

FIGURE XII-17: Power Supply With Variable Slope, Voltage And Inductance

One way to build the power supply shown in Fig. XII-17 is shown in Fig. XII-18.

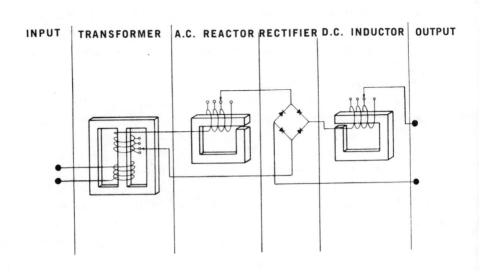

FIGURE XII-18: A Simple MIG Power Supply

138

The tapped transformer provides the variable voltage requirement. The tapped AC reactor gives variable slope, and the tapped inductor gives variable inductance. These *are not* the *only ways* to build such a power supply. There are a lot of ways to build power supplies (refer to Chapters VI, VII and VIII).

## Two Ways to Run an Arc

Running a MIG arc is a bit more complicated than running a TIG arc. The arc length in a TIG system depends on the operator and how long he holds the arc. In a MIG system the arc length depends on a balance between the melt rate and feed rate of the electrode. The only time the arc length is constant is when the melt rate and feed rate are equal, assuming, of course, that the torch to work distance remains constant.

There are two ways to control the length of a MIG arc. The first way is to hold the *melt rate constant,* by using a constant current type power supply and *vary the feed rate* of the electrode. By making the wire feed faster, or slower, the arc length can be made shorter or longer. This way requires a wire feeding system which can automatically adjust the feed rate to maintain the correct arc length.

The second way is to hold the *feed rate constant* and *change the melt rate,* up or down, as necessary. When the melt rate is increased, the arc gets longer, and vice versa. This way requires a power supply system which can automatically increase or decrease the current, in order to increase or decrease the melt rate.

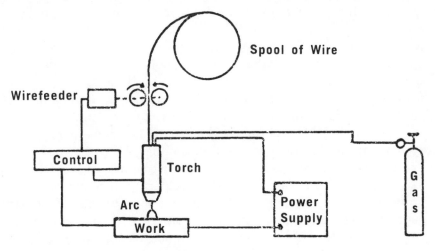

FIGURE XII-19: Typical MIG Welding System With Voltage Control

The first type, where the feed rate is varied, is called a VOLTAGE CONTROLLED system. It is shown in Fig. XII-19. The arc length is sensed by measuring the arc voltage. When the arc gets too long, as indicated by too much voltage, the feed rate is increased. When the voltage (length) is too short, the feed rate is decreased. The feed rate is changed automatically to hold a constant arc length. Fig. XII-19 is the same as Fig. XII-1, except for the voltage feedback control circuit.

The second type, where the feed rate is constant and the melt rate varies, is called the SELF REGULATING CONSTANT POTENTIAL system. It is shown in Fig. XII-1. Each type of system is discussed in the following paragraphs.

## The Voltage Controlled System

As was discussed earlier in this chapter, the electrode melt rate is proportional to the arc current. When the current is constant, the melt rate is constant. For all practical purposes, the current supplied by a "conventional" power supply is constant*. The slight increase or decrease in current (with a change in arc length) will not cause any significant change in the electrode melt rate.

When the feed rate is also constant, the arc cannot maintain its length. The arc length wanders all over the place and does one of two things. It either slowly increases until the arc burns back to the contact tube, or shortens until it stubs into the workpiece. These lengths are usually called the "burnback limit" and the "stubbing limit". The faster the wire feed system can respond, and the more sensitive the control, the easier it is to hold the arc length constant. A well designed voltage controlled wire feeding system virtually eliminates burnbacks and stubbing. However, the control will not prevent mishaps or occasional malfunctions in the wire feed system. Burnbacks do occur on occasion. Consequently, the current carrying contact tip is designed to be replaceable.

Although MIG welding is done with DC power, this type of system can be designed to work with either AC or DC power. For example, submerged arc welding can use either AC or DC. MIG arcs are usually operated in the SPRAY transfer mode when using voltage control. The limited short circuit current of constant current power supplies makes it impractical to use voltage controlled welding with the short arc mode of transfer.

The main thing to remember about the *voltage controlled*

* See Chapter VII, Fig. 4.

welding system is that the wire feed speed is changed in order to hold the arc voltage constant. The speed increases when the arc becomes too long, and decreases when the arc is too short. The wire feed speed is changed automatically by a control system which monitors the arc voltage, as shown in Fig. XII-19.

## The Self Regulating Constant Potential System

As was pointed out earlier in this chapter, the electrode melt rate can be considered to be proportional to the arc current. The melt rate increases when the arc current increases as is shown in Fig. XII-3. Twice as much current will melt approximately twice as much wire. Half as much current melts half as much wire, etc. etc. Fortunately, this is exactly what is needed to make a "self-regulating" system work.

The principle of self regulation can be best explained by way of an example. Imagine that a MIG arc, using 1/16" diameter aluminum wire, is operating at 300 amperes and 30 volts. There is no particular significance to these numbers, they were only chosen as a matter of convenience. It happens that for 1/16" diameter aluminum wire, that its melt rate is about 1 inch per minute per ampere. It is not 10 IPM/ampere or 0.1 IPM/ampere, but about one IPM/ampere. Therefore, in this imaginary example, the melt rate is 300 inches per minute because three hundred amperes will melt 300 IPM of 1/16" diameter aluminum wire. In order for the arc length to remain constant, the feed rate must also be 300 IPM. If the feed rate did not equal the melt rate, the arc length would vary, and the arc could either "burnback" or "stub out". Finally, assume that the arc current is supplied by a constant potential

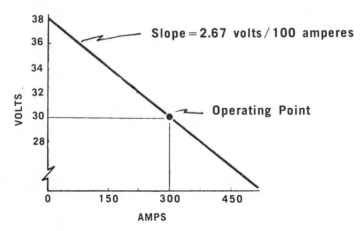

FIGURE XII-20: Typical Steady State MIG Operating Point

power supply system with a built in slope of 2.67 volts/100 amperes*. The slope includes the resistance of the leads etc. At 300 amperes, the power supply voltage is 8 volts less than its open circuit voltage. $(3 \times 2.67 = 8)$ Therefore, the open circuit voltage is 38 volts. The numbers above are summarized in Figure XII-20.

The operating point in Fig. XII-20 is shown as a dot on the volt/ampere curve of the power supply. The V/A curve starts out with 38 volts at zero current and slants downhill at 2.67 volts/100 amperes. At 300 amperes, the power supply puts out 30 volts. As long as the feed rate and melt rate are both equal, and the torch to workpiece distance does not change, the operating point will remain in the same place. Self-regulation of arc length does not occur under the conditions described so far.

Self regulation comes into play when the arc length is changed. Whenever the torch is raised or lowered, whenever the workpiece changes height, or whenever the weld pool moves the arc length tends to change. Assume for a moment that the arc becomes longer, about 2 volts longer. In other words, the arc voltage becomes 32 volts, instead of 30 volts. This requires the operating point to move upwards, and to the left, along the power supply V/A curve shown in Fig. XII-20. When the operating point is at 32 volts, the current is only 225 amperes**. The new operating point is shown in Fig. XII-21a.

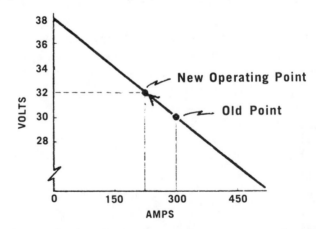

FIGURE XII-21a: Shift In Operating Point Due to Length Increase

The only place on the V/A curve that can provide 32 volts is the 225 ampere point. With 225 amperes of current, instead of 300

* Once again, there is no particular significance to these numbers, except that they are representative of actual systems.

**Since the slope is 2.67v/100 amperes, 75 amperes will cause a 2 volt change.

amperes, the electrode melt rate falls off to 225 IPM. The original melt rate was 300 IPM, it is now 225 IPM, or 75 IPM less. The feed rate is still 300 inches per minute. The feed rate is now 75 IPM higher than the melt rate. The arc begins to shorten its length. The correction starts out at a rate of 75 IPM and decreases as the arc becomes shorter. As long as there is more *feed rate* than

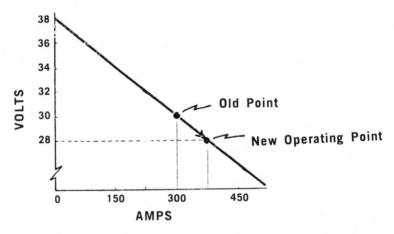

FIGURE XII-21b: Shift in Operating Point Due To A Length Decrease

*melt rate,* the arc length will shorten. When the arc is back to its 30 volts length, the current is 300 amperes. Once again the melt rate is 300 IPM (to go with 300 amperes) and the arc length is at its original length.

In the example just discussed, if the arc had become shorter (instead of longer) by two volts, the current would have increased to 375 amperes. The operating point on the power supply curve would have been 28 volts and 375 amperes, as shown in Fig. XII-21b. In this case, the extra 75 amperes melts an extra 75 IPM of electrode. Since the wire feed speed remained at 300 IPM, the extra 75 IPM of melt rate causes the arc to become longer. When the arc reaches its 30 volts length, the current is 300 amperes, and once again the melt rate and feed rate are equal.

No matter which direction the arc length is changed, either increased or decreased, the self-regulation effect tries to drive the arc back to its original length. The self-regulation effect depends on the increase or decrease of the electrode melt rate. The melt rate automatically increases when the length is decreased, and vice versa. The only way to produce a *permanent* change in arc length

143

is by changing the feed rate. When the feed rate is increased, the arc length becomes shorter and the current rises. The melt rate increases as the current increases until the feed rate and melt rate are balanced. The opposite occurs when the feed rate is decreased. Then the arc length increases and the current decreases. The melt rate follows the current and also decreases. When the melt rate becomes equal to the new feed rate the system is balanced.

The self-regulation effect automatically adjusts the melt rate to equal feed rate. The speed of *length correction* depends on the size of the current swing which is caused by length changes. The greater the current swing, the greater the speed of correction. With *no* current *swing,* there is *no correction.* That is why "self regulation" does not work when *constant current* power supplies are used for MIG welding*.

The speed of correction can be increased by using a power supply with less slope. For instance, if the slope of the power supply in the previous example was cut in half, the size of current swing would be doubled and the speed of correction would also be doubled. This is illustrated in Fig. XII-22. As shown, the current swing doubles to 150 amperes instead of 75 amperes, when the slope is cut in half. The dashed line, representing a power supply with a 1.33 volt/100 ampere slope, is marked *fast* in comparison to the original power supply V/A curve which is marked *slow*.

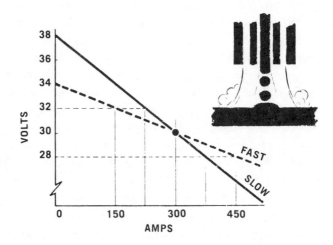

**FIGURE XII-22: Influence of Slope on Speed of Self Regulation of Arc Length**

* On occasion, particularly when a conventional CC supply has a high current setting, a CC supply can have enough of a current change to produce a useful self-regulation effect.

The actual time of correction is usually too short for the welding operator to notice the change visually. The difference in rate of adjustment between the two curves shown in Fig. XII-22 will not be seen by most operators. As an example of the short time of correction, examine the following illustration. Assume there is an average difference of 60 inches per minute between the melt rate and feed rate. Therefore, in *one second* the arc would change *one inch*. A step change of 0.100 inches would only need 0.1 seconds to "self regulate" to its original length.

The thing to remember about self regulation is that it works and that it works fast. It works best with constant potential type power supplies because they provide large current changes when the arc length is changed.

# Chapter XIII

# SHORT CIRCUITING TRANSFER
# MIG WELDING POWER SUPPLIES
## (Short Arc Power Supplies)

### Introduction

Although short circuiting transfer MIG welding was mentioned in Chapter XII, it is being discussed again in this chapter. The process has a number of distinctive characteristics which are described in detail in the following pages.

The term "short circuiting transfer MIG welding" is cumbersome. Many manufacturers of welding equipment have coined their own terms to describe the process. These terms include "Micro-wire gas metal arc" welding, "Dip transfer" welding and "Short arc" welding.* The Russians and Japanese use terms which translate as "short circuit arc" welding and "short arc" welding. This text uses the term "short arc". Although it is not necessary, it would be helpful for the reader to review Chapter XII before he reads this chapter.

Short arc is a direct current welding process used for joining materials up to several inches thick, as well as thin gauge materials. Reverse polarity is used for most applications. Short arc is particularly useful for welding steels of all types. Aluminum can be welded with short arc, but the results are not particularly good because short arc is a relatively "low heat" process. The *high* thermal conductivity of *aluminum* causes the puddle to freeze rapidly. The rapid freezing action traps gases in the weld puddle and creates porosity. The *low* thermal conductivity of *steel* delays the weld puddle solidification long enough for the gas bubbles to rise to the surface and escape.

One of short arcs most interesting features is its ability to bridge gaps and weld where joint fitup is poor. Its low heat input and relatively fast puddle freezing (compared to spray arc MIG welding) makes short arc very useful for welding in all positions. When needed, the heat of the arc can be increased by using special shielding gas mixtures and by "tuning in" the power supply. This chapter discusses short arc in general terms in order to aid the reader in "tuning in" his power supply.

### General Characteristics

Short arc is characterized by repeated short circuiting of the

* Other terms that have been used are "Closarc" welding and "Pinch arc" welding.

electrode and workpiece. When the electrode tip touches the weld puddle, the arc goes out. A liquid metal bridge is formed between the electrode tip and the puddle each time the tip touches the puddle. The same thing happens when you put one of your fingers into a glass of water. As soon as the finger touches the surface of the water, the water wets the end of your finger and forms a liquid bridge, as shown in Fig. XIII-1.

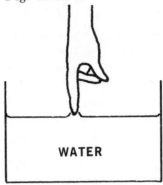

WATER

FIGURE XIII-1: Example of Liquid Bridge Formed by Surface Tension

There is one minor difference between the liquid bridge of the "finger-water" illustration and the short arc case. In short arc, the molten metal on the end of the electrode becomes part of the bridge. Part of the molten electrode tip is transferred to the weld puddle each time the bridge is pinched off*. The drops which are transferred to the puddle are very small. It takes a great many drops per second to supply enough metal to make a weld bead. For example, assume that each drop is 1/32'' in diameter, and that there are 120 drops transferred per second. This is equivalent to a deposit rate for steel of about 2 pounds an hour. If the wire electrode was also 1/32'' in diameter, the electrode feed rate would be approximately 150 inches/minute.

The important thing to remember about short arc is that metal is only transferred when the electrode touches the puddle. The welding action is completed by the arc, during the time between short circuits.

## Short-Circuit Frequency

The average frequency of short circuits (metal transfers) will vary from 20 to 200 times a second, depending on welding conditions. Whatever the frequency happens to be, at any one time, it

* See Chapter XII.

is not exactly constant. The normal differences in materials, equipment and power causes the frequency to vary from one second to another. For example, there might be 108 shorts for one second and then 132 shorts for the next second, etc. etc. Over a period of a minute or so, the average frequency would be 120 shorts per second. At some other current and wire feed speed, the average frequency might be 70 shorts per second, etc. etc.

The relatively high frequency of short arc (above 20 transfers per second) *cannot* be seen by the operator. Although high speed motion pictures will show that the arc is going *off* and *on,* the operator does not see it. The arc looks as if it is *on* all the time. An oscillograph record of the arc voltage and current is another way to show that the arc goes *off* and *on.* Figure XIII-2 shows a typical current and voltage record.

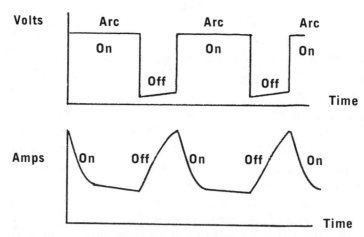

FIGURE XIII-2: Typical Current & Voltage Short Arc Oscillogram

The slightly random nature of the short circuit frequency requires the use of an *average* frequency for practical measurements. The frequency (f) can be expressed by relating it to the electrode feed rate (F) and the length of electrode (z) transferred with each drop. This is shown in equation XIII-a:

(XIII-a) $$f = \frac{F}{z} \qquad \text{Frequency} = \frac{\text{feed rate}}{\text{length}}$$

For example, with a feed rate of 360 inches per minute (6 inches per second), and an average length of 030'' of electrode used per transfer, the frequency is 200 shorts per second. This is illustrated in Fig. XIII-3:

$$\text{Frequency} = \text{Feed rate/length}$$
$$f = F/z$$

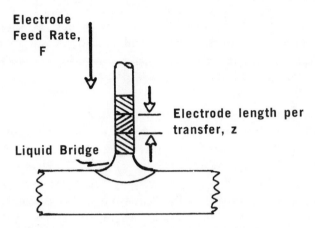

FIGURE XIII-3: Definition of Short Arc Transfer Frequency

## The Key to Stability

In order to get the most stable short arc metal transfer, disturbances of the cycle must be minimized. The smoothest operation occurs when the arc length change, caused by the pinch action, is kept to a minimum. This is so because, when the length change is small, the cycle is least disturbed and changes in arc power are kept to a minimum. The trick to keeping the arc length change small is to make the size of the transferred metal drops as small as possible. In terms of over-all stability, small drops are better than large drops. This means that the average length of wire (z) used per transfer should be as small as possible. This happens when the ratio of Feed rate to frequency (F/f) is small, as shown in equation XIII-b) which is a rearrangement of equation XIII-a.

(XIII-b)  $z = \dfrac{F}{f} = \dfrac{\text{Feed rate}}{\text{frequency}} =$ Electrode length used per transfer

In other words, the smaller the ratio F/f, the more stable the arc. Generally, the feed rate is fixed in advance. When the feed rate is constant, the frequency determines the ratio. Obviously, the frequency must be high to get a small ratio and a stable arc. Control of the frequency is the key to keeping the ratio of (F/f) small.

The KEY TO ARC STABILITY IS KEEPING THE RATIO OF (F/f) AS SMALL AS PRACTICAL.

One note of caution is worth mentioning. Just because the arc is most stable when the ratio is small does not mean that this is necessarily the best welding condition. Other things such as poor fit up may require an operator to use a condition which has a larger ratio. The following paragraphs discuss the effects of the power supply on arc stability and frequency.

### Open Circuit Voltage (OCV) and Stability

Power supply open circuit voltage (OCV) is one of the most important factors in controlling short arc frequency. The OCV controls the *arc voltage* and *length* by raising or lowering the operating point of the arc-power supply system.* As long as the power supply slope is fixed the OCV controls arc length, and, consequently, the frequency. The exact effect of arc voltage on frequency is shown in Fig. XIII-4.

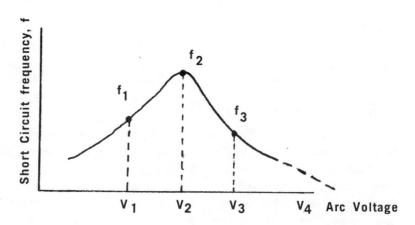

FIGURE XIII-4: Effect of ARC Voltage on Short ARC Frequency

The data for plotting the Frequency vs. Voltage curve is obtained by counting short circuits on an oscillogram or with a special short circuit counter. The voltage is obtained via a meter connected across the arc. The "hump" shape of the curve in Fig. XIII-4 can be explained as follows:

* See Chapter XII.

1) With *low* voltage (V₁) the arc stumbles because of too little power and causes an erratic *low* frequency (f₁).

2) With *medium* voltage (V₂) the arc stumble is eliminated and the frequency rises to a *high* value (f₂).

3) With *high* voltage (V₃) the arc becomes longer and allows large drops to form; large drops use more wire per transfer and the frequency decreases (f₃).

4) With *too much* voltage (V₄) the arc gap is too large for shorts to occur at all. Metal is transferred without shorts and the frequency is zero.

The highest frequency on the curve gives the best stability because it gives the smallest (F/f) ratio. The curve will shift its position with changes in feed rate, wire diameter, gas etc. However, there will always be a peak frequency point on the curve. Most operators use conditions near the peak point. Although they do not use an oscillograph to count shorts, or frequency meters, their eyes and ears prefer arc conditions that are near the peak of the curve. They usually adjust the arc voltage up and down until they find a condition they like. The condition they pick is always near, or at, the peak voltage-frequency point.

### Slope (V/A) and Stability

When the volt/ampere slope of the power source is changed, the arc voltage changes. As the slope is increased, the arc voltage decreases and vice versa. For any particular set of welding conditions (feed rate, OCV, gas, wire etc.) changes in slope cause changes in arc voltage and in frequency. The discussion about Fig. XIII-4 applies to slope (V/A) as well as open circuit voltage (OCV).

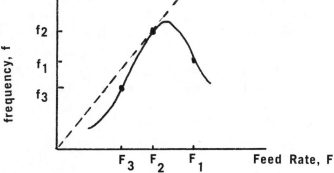

FIGURE XIII-5: Effect of Feed Rate on Frequency

151

## Feed Rate (F) and Stability

Changes in the feed rate cause changes in short arc frequency, when all other things remain constant. The general effect of feed rate on frequency is illustrated in Fig. XIII-5.

The shape of the curve can be explained as follows:

1) With *low feed rates* ($F_3$), the wire is fed too slowly, the arc becomes long. Large drops form causing the *frequency* to be *low* ($f_3$) or even zero.

2) With *medium feed rates* ($F_2$), the arc becomes shorter, the current rises and small drops are formed. The *frequency* rises ($f_2$).

3) With *high feed rates* ($F_1$), the wire is fed too fast. The arc becomes too short and the arc begins to stumble. The *frequency* becomes erratic and *lower* ($f_1$).

It is obvious that there is a higher frequency between $f_2$ and $f_1$. The peak frequency on the curve is higher than either one. The obvious conclusion *might* be that the highest frequency would be the best operating point. In this case, the answer would be wrong. The highest frequency *is not* the best frequency. The best frequency is at the point marked ($f_2$). Here is why. Remember from equation XIII-b that the key to stability is keeping the ratio of Feed rate to frequency (F/f) small as possible? Since Figure XIII-5 shows a plot of *both* feed rate and frequency, a slanted line drawn from the zero point to each point on the curve represents the ratio of *frequency* to *Feed rate* (f/F)*. The biggest ratio of frequency (f) to feed rate (F) is at the point marked by ($f_2$). It is the point where the slanted line is tangent to the curve. The stability ratio is the reverse of this ratio. In other words, the reciprocal of this ratio is the stability ratio. Therefore, in this case the largest (f/F) ratio is the same as the smallest (F/f) ratio. The point marked ($f_2$) has the smallest (F/f) ratio and represents the most stable operating point according to equation XIII-b.

When an operator adjusts the feed rate up and down, as he "tunes in" the welding condition, he usually picks a feed rate near the $f_2$ point. He usually likes that condition better than all the others. The important thing to remember here is that the *best* condition does *not* go with the *highest* frequency.

## Inductance and Stability

Inductance does several useful things for short arc. One of

---

* Slant = rise/run = frequency/feed rate.

which is to help "tune in" the best frequency.

When the inductance is tuned properly, it is possible to obtain the optimum (F/f) ratio. Changes in inductance cause a shift in the position of the Feed rate vs. frequency curve of Fig. XIII-5. This effect is illustrated in Fig. XIII-6, which shows the F vs. f curves for several values of inductance.

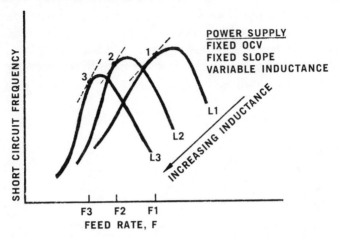

FIGURE XIII-6: Effect of increasing inductance on short-circuit frequency, f, at selected feed rates, F.

Note that as the inductance is increased the frequency curve moves to a lower position. As a general rule "increasing inductance decreases frequency". The explanation is complicated and beyond the scope of this book (it has to do with the effect of inductance on the arc voltage and arc *on* time, after a short circuit is pinched off).

As discussed under "Feed rate and Stability", the highest (F/f) ratio is the most stable operation condition of all the available conditions. In the same way, the best value of inductance *does not* corresponds to the highest frequency. The best choice of inductance is the one which gives the largest (f/F) ratio, and consequently the smallest (F/f) ratio.

The operator does not know what (F/f) ratio he is using, but the condition he selects as best usually uses an inductance of near the optimum value. For each particular feed rate there is a best value of inductance. This suggests that a power source with a variable inductance is desired to obtain the smoothest transfer over a wide range of feed rates and materials.

153

## Tuning in

As has been pointed out, there is an optimum arc voltage, which can be selected by changing the OCV or slope, or both in combination. For any particular power source setting, there is an optimum feed rate. The best feed rate is not the one which gives the highest frequency but gives the best (F/f) ratio. There is an optimum inductance value as well. When all the variables are juggled properly, it is possible to tune into the optimum condition. However, there is an easier way to "tune in".

As shown in Fig. XIII-7, it is possible to determine a zone of good welding conditions on a graph of *arc volts vs. feed rate.*

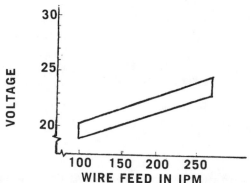

FIGURE XIII-7: Typical Zone of Good Short Arc Welding Conditions

The zone of good short arc conditions is about two volts wide. Whenever an arc voltage-feed rate combination falls inside the operating zone a good short arc condition is obtained. There are two ways to gather the data for operating zones as shown in Fig. XIII-7.

One way is to combine the data of arc voltage vs. frequency as shown in Fig. XIII-4 with the data of feed rate vs. frequency of Fig. XIII-5. By eliminating the frequency data it is possible to replot the arc voltage and feed rate data for the optimum welding conditions.

Another way is to measure the voltage and feed rate used by an experienced operator. The welding conditions he prefers are quite often the same ones as derived by an Engineer who counts short circuits on an oscillogram.

It is also possible to plot the data as *arc voltage vs. arc current*. This is simply done by replacing the feed rate data with arc current found from an electrode burn off curve. When you do not have curves as shown in Fig. XIII-7 and you are not a skilled operator, what do you do then?

One way to get a good (stable) condition is to increase the wire feed speed until the arc begins to "bump" and "stub out" on the workpiece. *Then decrease* the feed rate slightly until the erratic bumpy action disappears. The frequency ratio will be near the optimum.

Instead of changing the wire feed speed it is also possible to change the power source output. When the slope is set (fixed at one value), merely decrease the voltage setting until the arc begins to *bump* and *stub out*. Then increase the voltage approximately 2-3 volts. This method, too, will give a good frequency ratio. A similar effect can be obtained with slope controlled power supplies. When the slope is increased (steeper), the arc will begin to stumble and stub out. Decreasing the slope (flatter) will eliminate the stumbling. Avoid taking out too much slope, otherwise, the spatter percent will increase.

Most short arc welding is done with a standard MIG welding set up. The wire feed speed (current) is adjusted by means of separate "knob" on the wire feed speed control box. The arc voltage (length) is set by adjustment of a separate "knob" on the power supply. Welding conditions are readily selected by "tuning in" both of the settings.

There is even a short arc welding system which operates with "one knob" instead of two. The wire feed and the power supply adjustment knob mechanisms are interconnected mechanically. Adjustment of the "one knob" automatically selects the feed rate and arc voltage to place the operating point inside the zone shown in Fig. XIII-7.

# CHAPTER XIV

# FLUX-SHIELDED ARC WELDING POWER SUPPLIES

## Introduction

There are a number of welding processes similar to the MIG process which use a flux instead of, or in conjunction with, a shielding gas. In addition to stabilizing the arc, the flux is used to clean the metal and protect the molten metal from contamination. It is even possible to control the weld puddle composition to some extent by use of different fluxes.

Flux can be added to the weld zone in several ways. One way is to coat the welding electrode with flux, as is done with Covered Electrode Welding*. Another welding method uses a wire with the flux inside. This is called flux cored wire welding. A third way is to sprinkle a thick layer of flux over the weld zone. This is called submerged arc welding. Other variations use gas borne flux, magnetic flux, and various combinations of gases and fluxes. This chapter discusses two variations of the flux shielded welding processes. These are the submerged arc process and the cored wire process. These two processes are typical of all the flux shielded processes.

## The Submerged Arc System

The "sub arc" system, as it is often called, is very much like a MIG system. It can be used in either the voltage controlled constant current method or the self regulating constant potential method**. The typical CP system is shown in Fig. XIV-1. Most

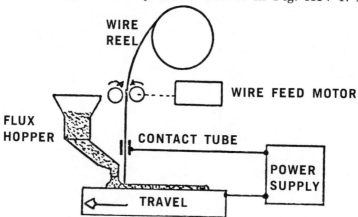

**FIGURE XIV-1: Typical CP Submerged Arc System**

\* See Chapter XI
\*\* See Chapter XII

submerged arc set-ups use mechanized equipment to move the arc along the weld seam.

The electrode is fed at a constant rate thru the flux layer. The heat in the arc zone melts the flux. The arc itself is surrounded by molten flux which solidifies after the arc passes. The unused flux may be removed and reused. The fused flux usually peels off of the weld as it cools. This "self peeling", or "free peeling" characteristic is very desirable because it reduces the clean up work. The arc current is controlled by increasing or decreasing the electrode feed rate and the arc length is controlled by changing the voltage output of the power supply. Fig. XIV-2 illustrates what is happening beneath the blanket of flux.

Spatter is no problem because the arc is completely submerged beneath the layer of flux. The molten slag surrounding the arc traps the spatter, and because spatter is heavier than the molten slag it sinks into the weld pool. Since spatter is not a problem it is possible to use currents 2 to 3 times as high as used with MIG arcs. It is not unusual to have currents as high as 1500 amperes on a single electrode. The weld puddles are big and the power liberated in the arc zone is high. The sub arc process is ideal for use on heavy steel plate. It can also be used on thin materials, but its prime use is on heavy steel weldments.

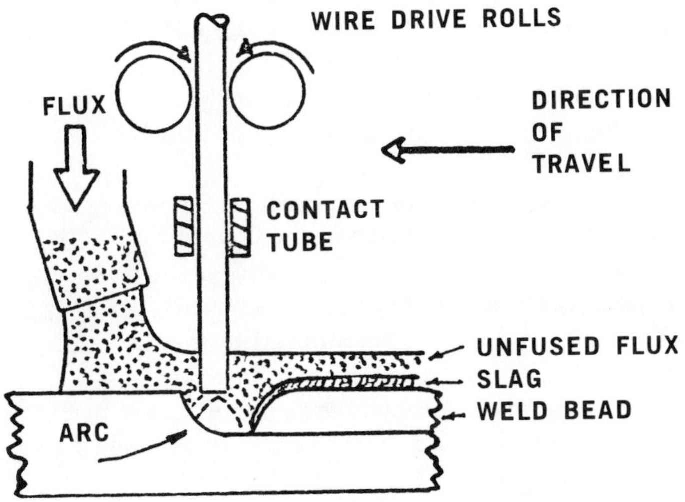

FIGURE XIV-2: Cross Section of Submerged Arc Zone

157

In addition to holding down the spatter level, the flux makes it easier to run an arc. Gases liberated in the weld zone make it possible to run AC arcs as well as DC arcs. The various types of sub arc power are discussed in the following paragraphs.

## Constant Potential DC Submerged Arc Power

Because spatter is no great problem in sub arc welding, there is no need to use slope for controlling the short circuit current. Power supplies without slope are less expensive then those with slope. Therefore, most direct current constant potential sub arc power supplies have very flat volt/ampere curves. A large amount of inductance is *recommended* for smoothing out the large current swings which would occur if it were not present. Large current swings tend to cause puddle turbulence. Inductance helps to quiet the arc action. However, inductance is not absolutely essential in running submerged arcs.

A CP submerged arc is usually started in one of two ways. One way is to run the electrode into the plate at welding speed and let the short circuit current surge fuse off the wire, just as in MIG welding. In this type of start, the current can rise to several thousands of amperes. Another way is to start the arc is called a "flying start". In this method the carriage, which moves the wire feed mechanism along the weld seam, is turned on *before* the electrode wire feed motor is energized. Then, when the electrode runs into the weld zone the motion of the carriage causes the electrode to scratch the workpiece. The scratching action helps start the arc. The static and dynamic requirements of a CP sub arc power supply are not very complicated. Practically anything which can provide *enough* current and voltage will work. This is in contrast to MIG welding which requires specific static and dynamic characteristics.

## Constant Current Voltage Controlled Submerged Arc Power

This method of submerged arc welding is like MIG arc welding with CC power. The arc length (voltage) is maintained by variations in the wire feed speed. A reference arc length (voltage) set on the welding control is compared to the arc voltage. The feed rate adjusts itself automatically to maintain the arc voltage the same as the reference voltage. The current is set by adjustment of the power supply. Fig. XIV-3 illustrates a typical CC submerged arc welding system.

There are several ways to start the arc with this method of sub arc welding. One way is called "retract starting". With this

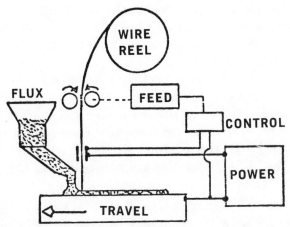

FIGURE XIV-3: Typical CC Voltage Controlled Sub Arc System

method, the electrode is inched down into the work. As soon as current flow is sensed the wire is *retracted off* the work. As the wire retracts, it breaks contact with the work. The retract action is like opening a switch when current is flowing. When the weld circuit is broken by the retract action a spark occurs in the arc gap. The spark grows into a welding arc. As soon as the arc is started, the wire feed direction is changed automatically to feed *toward* the workpiece and normal welding begins. Another way to start a sub arc is called a "steel wool start". In this method, a small piece of steel wool is rolled into a ball. The ball of steel wool is placed on a clean spot on the workpiece. Then the electrode is inched down (without power) until it just touches the steel wool ball. The weld zone is covered with flux and then simultaneously the power sup-

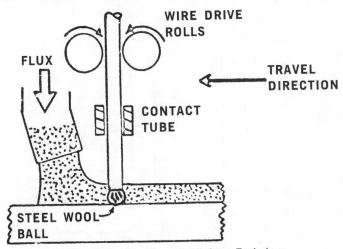

FIGURE XIV-4: The "Fuse Ball" Start Technique

159

ply is turned on and the wire feed is started. Welding current flows through the steel wool ball and "fuses" it away. The arc starts in the space left by the melted steel wool. A steel wool start is illustrated in Fig. XIII-4.

A third, very common way, is to use the "flying start" technique discussed in the previous topic on CP power. The scratching motion of the electrode on the workpiece helps start the arc. A rather rigid mechanical carrier system is required for this type of CC start. Otherwise the feeding electrode will merely lift the carriage instead of starting the arc by scratching the workpiece.

## AC or DC?

Alternating current is preferred whenever magnetism in the steel causes "arc blow". Arc blow is caused by interaction of the magnetic field of the arc and magnetism in the weld plate. It causes erratic metal transfer and an irregular bead shape. When DC is used the arc is deflected in one direction, with AC the deflection changes when the current changes direction. The alternating arc deflection tends to even out the arc blow effects. In addition, the AC field of the arc tends to neutralize the magnetic field in the weld plate.

When two DC arcs are operated in the same arc zone their magnetic fields interact. Each of the DC arcs causes the other to move. They produce mutual arc blow. When two AC arcs, of the same phase relationship, are operated in the same arc zone they also cause mutual arc blow. However, with AC arcs, it is possible to use *phase shift** to minimize (or control) arc blow. When one of the two AC arcs is at its maximum magnetic strength (maximum

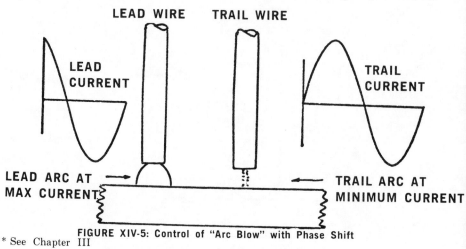

FIGURE XIV-5: Control of "Arc Blow" with Phase Shift

* See Chapter III

160

current) the other one can be at zero. This is illustrated in Fig. XIV-5. The thing to remember about AC sub arc power is that it is used to control arc blow.

## AC Sub Arc Stability

Alternating current arcs extinguish each time the sine wave passes thru zero, just as in AC TIG welding*. An AC TIG arc requires the use of High Frequency power to reignite the arc each time it goes out. An AC sub arc does not need HF power. The gases liberated by the flux makes it easy for a CC power supply to reignite the arc. The open circuit voltage normally available is sufficient to produce reignition. The normal OCV for conventional AC sub arc power supplies is about 80 volts.

Constant potential AC sub arc power is not normally used. It is difficult to produce good arc reignition with the lower OCV of CP AC power supplies.

## The Two Wire Scott Connection

One method of multiple wire submerged arc welding uses the so called Scott Transformer Connection. It is named after C. F. Scott, the inventor of the transformer connection. The Scott Connection was developed to change normal three phase AC (3∅ AC) into two phase AC (2∅ AC), or vice versa. Submerged arc welding uses the *three phase into two phase* version of the system.

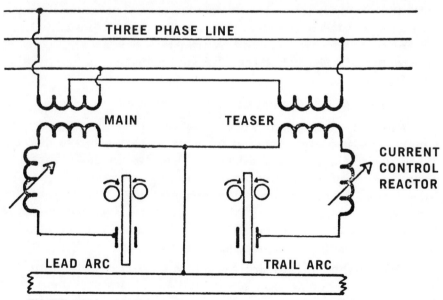

FIGURE XIV-6: The Two Electrode Scott (2 Phase) Connection

* See Chapter IX

With two phases, one for each of two electrodes, arc blow is held to a minimum. The two phases are 90° to each other. When one is at maximum, the other is at minimum and vice versa. Not only that, the phase currents can be adjusted separately without effecting each other. One phase is usually called the *lead* electrode phase and the other is called the *trail* electrode phase. The two electrodes operate one behind the other, in the same weld pool. The ability to separately adjust the current in each electrode is a great advantage. It simplifies adjustment of the sub arc system.

Fig. XIV-6 illustrates the Scott connected sub arc system. The reactor in each of the welding circuits is used to adjust the arc currents. The reactor in the lead electrode circuit adjusts the lead current, and the reactor in the trail electrode circuit adjusts the trail current. As long as the phase shift between the lead and trail currents is 90° there is no interaction between the current adjustments. In some *practical* Scott connections the phases are not at exactly 90°. In such pseudo-Scott cases there will be an interaction between the current adjustments. The further away the phase shift is from 90°, the more pronounced the interaction will become. As shown in Fig. XIV-6, two single phase transformers are connected to a three phase line. One of the transformers has its primary winding tapped at 50%. The 50% tap is connected to one terminal of the second transformer. The first (transformer with the 50% tap) is called the MAIN transformer. The second transformer (connected to the main transformers 50% tap) is called the TEASER transformer. The MAIN transformer is usually used to operate the LEAD arc, and the TEASER transformer operates the TRAIL arc.

### The Series Arc Submerged Arc Connection

With this connection two arcs are operated from one power supply. The arcs are connected in series as shown in Fig. XIV-7. When the arcs are separated by several inches the current travels thru the workpiece to complete the circuit. When the arcs are placed close to each other the current can flow from one to the other without entering the workpiece. As normally used this connection provides a high deposition rate with low penetration. It is an effective method of surfacing and cladding.

This method of sub arc welding has several disadvantages. For example, if one arc becomes unstable it causes the other arc to become unstable, because both arcs use the same current. In addition, it is difficult to attain reliable reignition of the arcs each

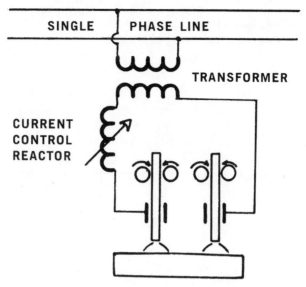

FIGURE XIV-7: Typical Series Arc Connection

time the sine wave passes thru zero. Two arcs require more voltage for reignition than one arc. Since the power supply is normally designed for use with one arc it may not have enough open circuit voltage for two arcs in series.

### Flux Cored Wire Welding

With this method of welding, a tubular wire is filled with flux and used in the same manner as a solid wire is with MIG welding. The flux acts to clean the metal and protect the weld from atmospheric contamination. Most flux cored wire welding is done with direct current, however, AC power can be used with the proper flux mixture. Arc stabilizers are mixed into the flux core in order to aid arc reignition. Auxiliary gas shielding is used in many cases to improve arc stability.

This is an open arc process. The flux does not surpress the spatter as it does in sub arc welding. The same power supplies which are used for MIG spray arc welding may be used for flux cored wire welding. However, because of the arc stabilizers in the flux, there are only a few short circuits during welding. It is not necessary to use slope for control of spatter, or inductance either. The arc will work quite well with a constant potential power that has little or no slope. It will also work with voltage controlled constant current welding equipment. As with sub arc, virtually any power supply with the correct amount of voltage and current will work. Some benefit will occur when using power supplies designed for MIG welding, but the flux cored wire process is not critical in terms of the type of power supply used.

# APPENDIX

TABLE I

## Resistivity (r) of Metals and Alloys at 20°C (68°F)

| Materials | Microhm-centimeters | Microhm-inches |
|---|---|---|
| Aluminum | 2.828 | 1.113 |
| Copper | 1.724 | 0.679 |
| Iron | 10 | 3.937 |
| Nickel | 7.8 | 3.07 |
| STEELS: | | |
| Low Carbon (Range) | 10-12 | 3.93 - 4.72 |
| Silicon (Range) | 40-50 | 15.75 - 19.7 |
| Stainless (Range) | 60-75 | 23.6 - 29.5 |

$$\text{Resistance} = \frac{r \times \text{length}}{\text{Cross section area*}}$$

## TABLE II
## OHM'S LAW

| When you know | USE THESE FORMULAS FOR FINDING THE UNKNOWN | | | |
|---|---|---|---|---|
| | Current I | Voltage E | Resistance R | Power W |
| Current and Resistance, I&R | | I R | | $I^2R$ |
| Current and Voltage, I&E | | | E/I | I E |
| Current and Power, I&W | | W/I | $W/I^2$ | |
| Resistance and Voltage, R&E | E/R | | | $E^2/R$ |
| Resistance and Power, R&W | $\sqrt{W/R}$ | $\sqrt{WR}$ | | |
| Voltage and Power, E&W | W/E | | $E^2W$ | |

* see Tables VII. VIII, and IX.

## TABLE III
### Power Factor (PF) Correction

| Original PF | Desired PF* | | |
|---|---|---|---|
| | 1.00 | 0.90 | 0.80 |
| 0.60 | 1.33 | 0.850 | 0.583 |
| 0.70 | 1.02 | 0.536 | 0.271 |
| 0.80 | 0.75 | 0.265 | 0 |
| 0.90 | 0.48 | 0 | 0 |
| 1.00 | 0 | 0 | 0 |

Use the factor in the table to find the capacitive voltamperes required for correction. The correction voltamperes = Load watts x factor. EXAMPLE: a power supply has a PF = 0.6 and a load of 10 KW, what capacitive voltamperes are required for raising the PF to 0.9? From the table find 0.85 as the factor, then the correction voltamperes are:

(10KW) x (0.85) = 8.5 KVA

ANSWER: 8.5KVA

## TABLE IV

The Ripple Factor for various forms of rectifiers is shown below:

| Type of rectifier circuit | Single Phase | | Three Phase | |
|---|---|---|---|---|
| | 1/2 Wave | Full Wave | 1/2 Wave | Full Wave |
| RIPPLE FACTOR | 1.21 | 0.48 | 0.21 | 0.04 |

$$\text{Ripple Factor} = \frac{\text{Effective (RMS) value of the AC components of the current (voltage)}}{\text{Average value of the current (voltage)}}$$

## Ripple Factor

The ripple factor is used to measure the *waviness* of a wave form. A smooth wave form (no bumps) has a ripple factor of zero. Ripple factor tells how much work a filter *has* to do, or *is* doing.

A detailed theoretical discussion of the Ripple Factor is beyond the scope of this book.

*PF is also referred to in %. A PF = 1 is the same as PF = 100%.

## TABLE V
### TUNGSTEN ELECTRODE CURRENT (IN AMPERES) RATINGS

Approximate Current Ranges for Tungsten and Thoriated Tungsten Electrodes*

| Electrode Diameter (in.) | Electrode Negative Direct Current — Pure Tungsten and Thoriated Tungsten | Electrode Positive Direct Current — Pure Tungsten and Thoriated Tungsten | Unbalanced Wave Alternating Current Pure Tungsten | Unbalanced Wave Alternating Current Thoriated Tungsten | Balanced Wave Alternating Current Pure Tungsten | Balanced Wave Alternating Current Thoriated Tungsten |
|---|---|---|---|---|---|---|
| 0.010 | — | — | — | — | — | — |
| 0.020 | 5-20 | — | 5-15 | 5-20 | 10-20 | 5-20 |
| 0.040 | 15-80 | — | 10-60 | 15-80 | 20-30 | 20-60 |
| 1/16 | 70-150 | 10-20 | 50-100 | 70-150 | 30-80 | 60-120 |
| 3/32 | 150-250 | 15-30 | 100-160 | 140-235 | 60-130 | 100-180 |
| 1/8 | 250-400 | 25-40 | 150-210 | 225-325 | 100-180 | 160-250 |
| 5/32 | 400-500 | 40-55 | 200-275 | 300-400 | 160-240 | 200-320 |
| 3/16 | 500-750 | 55-80 | 250-350 | 400-500 | 190-300 | 290-390 |
| 1/4 | 750-1000 | 80-125 | 325-450 | 500-630 | 250-400 | 340-525 |

* Using argon as the shielding gas.

## TABLE VI

### Approximate Spray Arc Transition Currents* (In Amperes)

| ELECTRODE DIA. | STEEL Argon+2% O2 | ALUMINUM Argon |
|---|---|---|
| .030″ | 155±5 | 90±5 |
| .035″ | 170±10 | 95±5 |
| .045″ | 220±10 | 120±10 |
| .0625 (1/16)″ | 275±20 | 170±15 |
| 0.094 (3/32″) | 370±25 | - - - - |
| 0.8 mm | 160±5 | 90±5 |
| 1.0 mm | 185±10 | 100±5 |
| 1.2 mm | 220±10 | 120±10 |
| 1.4 mm | 235±15 | 130±10 |
| 1.6 mm | 275±20 | 170±15 |
| 2.0 mm | 310±20 | - - - - |
| 2.4 mm | 370±25 | - - - - |

* The current varies with electrode extension, alloy content, shielding gas, etc.

# TABLE VII

## Wire Diameter in mm, Equivalent Diameter in inches, Area in Square Centimeters and Square Inches.

| Wire Diameter Millimeters (mm) | Equivalent Diameter Inches (in) | Area Square Centimeters (cm²) | Area Square Inches (in²) |
|---|---|---|---|
| 0.2 | .007874 | .0000314 | .0000487 |
| 0.4 | .015748 | .001257 | .0001947 |
| 0.6 | .023622 | .002827 | .0004381 |
| 0.8 | .031496 | .005027 | .0007790 |
| 1.0 | .03937 | .007854 | .001217 |
| 1.2 | .047244 | .01131 | .001753 |
| 1.4 | .055118 | .01539 | .002386 |
| 1.6 | .062992 | .02011 | .003116 |
| 1.8 | .070866 | .02545 | .003944 |
| 2.0 | .07874 | .03142 | .004869 |
| 2.4 | .094488 | .04524 | .007012 |
| 2.6 | .10236 | .05309 | .008229 |
| 2.8 | .11024 | .06158 | .009545 |
| 3.0 | .11811 | .07069 | .010956 |
| 3.2 | .12598 | .08042 | .012465 |
| 3.4 | .13386 | .09076 | .014073 |
| 3.6 | .14173 | .10179 | .015776 |
| 3.8 | .14961 | .11341 | .017580 |
| 4.0 | .15748 | .12566 | .019478 |
| 4.5 | .17717 | .15904 | .024653 |
| 5.0 | .19685 | .19635 | .030434 |
| 5.5 | .21654 | .23758 | .036827 |
| 6.0 | .23622 | .28274 | .043825 |

## TABLE VIII
### Wire Diameter in Inches, Equivalent Diameter in mm, Area in Square Inches and Square Centimeters

| Wire Diameter Inches | | Equivalent Wire Diameter Millimeters (mm) | Cross Section | |
| --- | --- | --- | --- | --- |
| | | | Square Inches (in$^2$) | Square Centimeters (cm$^2$) |
| | .01 | 0.254 | .00007854 | .0005067 |
| | .015 | 0.381 | .0001767 | .001140 |
| | .02 | 0.508 | .0003142 | .002027 |
| | .025 | 0.735 | .0004909 | .003167 |
| | .030 | 0.762 | .0007069 | .004560 |
| 1/32 | .03125 | 0.794 | .0007669 | .004951 |
| | .035 | 0.889 | .0009621 | .006207 |
| | .040 | 1.016 | .001257 | .008107 |
| | .045 | 1.143 | .001590 | .010261 |
| 3/64 | .046875 | 1.191 | .001726 | .011141 |
| | .05 | 1.270 | .001964 | .012668 |
| | .055 | 1.397 | .002376 | .015328 |
| | .06 | 1.524 | .002827 | .018241 |
| 1/16 | .06250 | 1.587 | .003068 | .019780 |
| | .07 | 1.778 | .003848 | .024829 |
| | .08 | 2.032 | .005027 | .032429 |
| | .09 | 2.286 | .006362 | .041043 |
| 3/32 | .09375 | 2.381 | .006903 | .044525 |
| | .10 | 2.540 | .007854 | .050671 |
| | .120 | 3.048 | .01131 | .072966 |
| 1/8 | .125 | 3.175 | .01227 | .079173 |
| 5/32 | .15625 | 3.969 | .01917 | .123723 |
| 3/16 | .1875 | 4.762 | .02761 | .178102 |
| 7/32 | .21875 | 5.556 | .03758 | .242446 |
| 1/4 | .250 | 6.350 | .04909 | .316692 |

**TABLE IX**
American And Metric Wire Gages,
Areas And Diameters

| AMERICAN (B & S) GAGE | METRIC GAGE | AT 20°C (68°F) | | | |
|---|---|---|---|---|---|
| | | DIAMETER | | AREA | |
| | | INCHES in. | MILLIMETERS mm | SQUARE INCHES in$^2$ | SQUARE CENTIMETERS cm$^2$ |
| 4/0 | | 0.460 | 11.68 | 0.1662 | 1.072 |
| 3/0 | | 0.4096 | 10.40 | 0.1318 | 0.8503 |
| | 100 | 0.3937 | 10.00 | 0.1217 | 0.7854 |
| 2/0 | | 0.3648 | 9.27 | 0.1045 | 0.6743 |
| | 90 | 0.3543 | 9.00 | 0.0986 | 0.6362 |
| 1/0 | | 0.3249 | 8.25 | 0.08289 | 0.5348 |
| | 80 | 0.315 | 8.00 | 0.07793 | 0.5027 |
| 1 | | 0.2893 | 7.348 | 0.06573 | 0.4241 |
| | 70 | 0.2756 | 7.00 | 0.05965 | 0.3849 |
| 2 | | 0.2576 | 6.544 | 0.05213 | 0.3363 |
| | 60 | 0.2362 | 6.00 | 0.04383 | 0.2827 |
| 3 | | 0.2294 | 5.827 | 0.04134 | 0.2667 |
| 4 | | 0.2043 | 5.189 | 0.03278 | 0.2115 |
| | 50 | 0.1968 | 5.00 | 0.03043 | 0.1963 |
| 6 | | 0.162 | 4.115 | .020620 | .1330 |
| | 40 | .1575 | 4.00 | .019480 | .1257 |
| 8 | | .1285 | 3.264 | .012970 | .08366 |
| | 30 | .1181 | 3.00 | .010960 | .07068 |
| 10 | | .1019 | 2.588 | .008155 | .05262 |
| 12 | | .08081 | 2.053 | .005129 | .03309 |
| | 20 | .07874 | 2.00 | .004869 | .03142 |
| | 18 | .07087 | 1.80 | .003944 | .02545 |
| 14 | | .06408 | 1.628 | .003225 | .02081 |
| | 16 | .06299 | 1.60 | .003116 | .02011 |
| | 14 | .05512 | 1.40 | .002386 | .01539 |
| 16 | | .05082 | 1.291 | .002028 | .01309 |
| | 12 | .04724 | 1.20 | .001753 | .00131 |
| 18 | | .04030 | 1.024 | .001276 | .008231 |
| | 10 | .03937 | 1.00 | .001217 | .007854 |
| 20 | | .03196 | 0.8118 | .0008023 | .005176 |
| | 8 | .03150 | .80 | .0007791 | .005027 |
| 22 | | .02535 | .6439 | .0005046 | .003255 |
| | 6 | .02362 | .60 | .0004383 | .002827 |
| 24 | | .02010 | .5106 | .0003173 | .002047 |
| 26 | | .01594 | .4049 | .0001996 | .001287 |
| | 4 | .01575 | .40 | .0001948 | .001257 |
| 28 | | .01264 | .3211 | .0001255 | .0008098 |
| 30 | | .01003 | .2548 | .00007894 | .0005093 |

**TABLE X**

Calculation of Conductor Average Temperature Rise
Based on NEMA Specification EW-1-1971

1. Measure in °C and Ohms*
   $T_{C1}$ = Temperature of room at start
   $T_{C2}$ = Temperature of room at finish
   $R_C$  = Resistance at start (cold)
   $R_H$  = Resistance at finish (hot)

2. Calculate for Copper

   Average Temperature Rise $= \dfrac{R_H - R_C}{R_C}(234.5 + T_{C1}) - (T_{C2} - T_{C1})$

3. Substitute 225 for 234.5 for calculation of temperature rise of aluminum conductors.


\* To convert from °F to °C:  $°C = \dfrac{5}{9}(°F - 32)$

173

## TABLE XI
## FEED RATE TO DEPOSITION RATE CONVERSION
## FACTORS FOR STEELS

in/min→lbs/hr

| Wire Diameter, Inches | To Convert Feed Rate in Inches/Minute To Deposition Rate in Pounds/Hour Multiply Inches/Minute by |
|:---:|:---:|
| .020 | 0.00533 |
| .030 | 0.012 |
| 1/32 | 0.013 |
| 045 | 0.027 |
| 3/64 | 0.0293 |
| 1/16 | 0.0521 |
| 3/32 | 0.1172 |
| 1/8 | 0.2083 |
| 5/32 | 0.3254 |
| 3/16 | 0.4686 |
| 7/32 | 0.6379 |
| 1/4 | 0.8332 |

$$\text{Factor} = \frac{16.974}{\text{in}^2} \times \text{(wire cross section in square inches)}$$

## TABLE XII
## FEED RATE TO DEPOSITION RATE
## CONVERSION FACTORS
## FOR STEEL
m/min→Kg/hr

| Wire Diameter, millimeters | To Convert Feed Rate in Meters/Minute To Deposition Rate in Kilograms/Hour Multiply Meters/Minute by |
|:---:|:---:|
| 0.2 | 0.001475 |
| .4 | .0590 |
| .6 | .1328 |
| .8 | .2360 |
| 1.0 | .3690 |
| 1.2 | .5313 |
| 1.4 | .7230 |
| 1.6 | .9448 |
| 2.0 | 1.476 |
| 2.4 | 2.125 |
| 2.8 | 2.893 |
| 3.0 | 3.321 |
| 3.2 | 3.778 |
| 4.0 | 5.904 |

$$\text{Factor} = \frac{46.98}{cm^2} \times (\text{wire cross section in square centimeters})$$

## TABLE XIII
## FEED RATE TO DEPOSITION RATE CONVERSION FACTORS FOR ALUMINUM

in/min→lbs/hr

| Wire Diameter, Inches | To Convert Feed Rate in Inches/Minute To Deposition Rate in Pounds/Hour Multiply Inches/Minute by |
|---|---|
| 020 | .001838 |
| 030 | .004135 |
| 1/32 | .004486 |
| 045 | .0093 |
| 3/64 | .0101 |
| 1/16 | .0179 |
| 3/32 | .0404 |
| 1/8 | .0718 |
| 5/32 | .1121 |
| 3/16 | .1615 |
| 7/32 | .2198 |
| 1/4 | .2872 |

$$\text{Factor} = \frac{5.85}{\text{in}^2} \times (\text{wire cross section in square inches})$$

## TABLE XIV
## FEED RATE TO DEPOSITION RATE
## CONVERSION FACTORS
## FOR ALUMINUM

$$m/min \rightarrow Kg/hr$$

| Wire Diameter, millimeters | To Convert Feed Rate in Meters/Minute<br>To Deposition Rate in Kilograms/Hour<br>Multiply Meters/Minute<br>by |
|:---:|:---:|
| 0.2 | .000508 |
| 0.4 | .02036 |
| 0.6 | .0458 |
| 0.8 | .0814 |
| 1.0 | .1272 |
| 1.2 | .1831 |
| 1.4 | .2492 |
| 1.6 | .3256 |
| 2.0 | .5088 |
| 2.4 | .7326 |
| 2.8 | .9972 |
| 3.0 | 1.1447 |
| 3.2 | 1.3023 |
| 4.0 | 2.0349 |

$$\text{Factor} = \frac{16.1934}{cm^2} \times \text{(wire cross section in square centimeters)}$$

# CHART I
## GAS FLOW CONVERSION CHART

$$ft^3/hr \rightleftharpoons 1/min$$

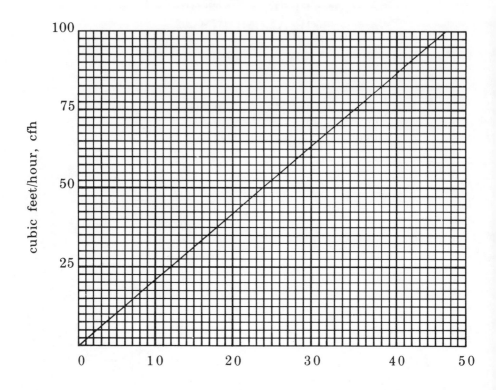

cubic feet/hour, cfh

liters/minute, lpm

One cubic foot per hour = 0.471948 liters/minute
One liter per minute = 2.118877 cubic feet/hour

# CHART II
## FEED RATE CONVERSION CHART

in/min $\rightleftarrows$ m/min

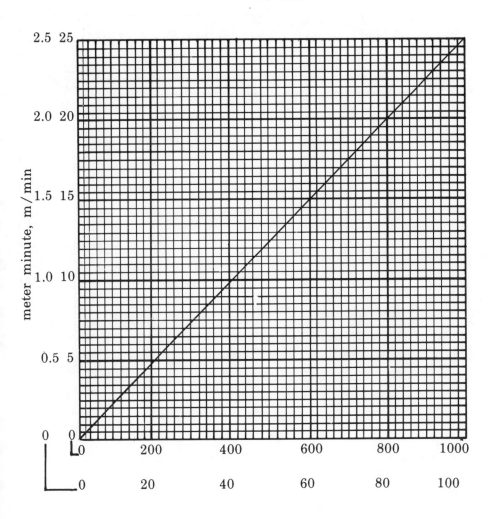

inches/minute, ipm

One inch/minute = .0254 meters/minute
One meter/minute = 39.37 inches/minute

## CHART III
## DEPOSITION RATE CONVERSION CHART

Kg/hr ⇌ lbs/hr

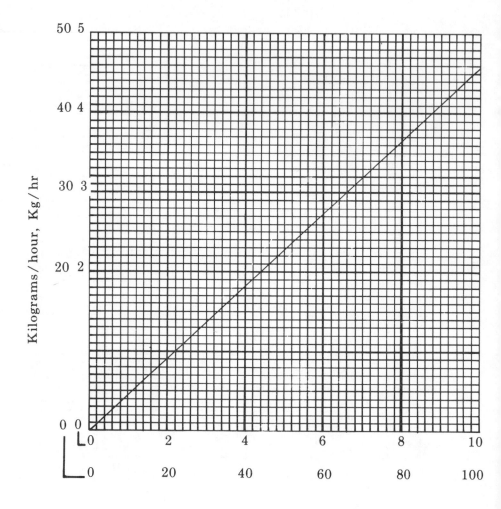

pounds/hour, lb/hr

One pound/hour = 0.4536 kilogram/hour
One kilogram/hour = 2.205 pounds/hour

180

# CONVERSION FACTOR TABLE*

| TO CONVERT | INTO | MULTIPLY BY |
|---|---|---|
| **A** | | |
| abcoulomb | Statcoulombs | $2.998 \times 10^{10}$ |
| acre | Sq. chain (Gunters) | 10 |
| acre | Rods | 160 |
| acre | Square links (Gunters) | $1 \times 10^5$ |
| acre | Hectare or sq. hectometer | .4047 |
| acres | sq feet | 43,560.0 |
| acres | sq meters | 4,047. |
| acres | sq miles | $1.562 \times 10^{-3}$ |
| acres | sq yards | 4,840. |
| acre-feet | cu feet | 43,560.0 |
| acre-feet | gallons | $3.259 \times 10^5$ |
| amperes/sq cm | amps/sq in. | 6.452 |
| amperes/sq cm | amps/sq meter | $10^4$ |
| amperes/sq in. | amps/sq cm | 0.1550 |
| amperes/sq in. | amps/sq meter | 1,550.0 |
| amperes/sq meter | amps/sq cm | $10^{-4}$ |
| amperes/sq meter | amps/sq in. | $6.452 \times 10^{-4}$ |
| ampere-hours | coulombs | 3,600.0 |
| ampere-hours | faradays | 0.03731 |
| ampere-turns | gilberts | 1.257 |
| ampere-turns/cm | amp-turns/in. | 2.540 |
| ampere-turns/cm | amp-turns/meter | 100.0 |
| ampere-turns/cm | gilberts/cm | 1.257 |
| ampere-turns/in. | amp-turns/cm | 0.3937 |
| ampere-turns/in. | amp-turns/meter | 39.37 |
| ampere-turns/in. | gilberts/cm | 0.4950 |
| ampere-turns/meter | amp/turns/cm | 0.01 |
| ampere-turns/meter | amp-turns/in. | 0.0254 |
| ampere-turns/meter | gilberts/cm | 0.01257 |
| angstrom unit | Inch | $3937 \times 10^{-9}$ |
| angstrom unit | Meter | $1 \times 10^{-10}$ |
| angstrom unit | Micron or (Mu) | $1 \times 10^{-4}$ |
| are | Acre (US) | .02471 |
| ares | sq. yards | 119.60 |
| ares | acres | 0.02471 |
| ares | sq meters | 100.0 |
| astronomical Unit | Kilometers | $1.495 \times 10^8$ |
| atmospheres | Ton/sq. inch | .007348 |
| atmospheres | cms of mercury | 76.0 |
| atmospheres | ft of water (at 4°C) | 33.90 |
| atmospheres | in. of mercury (at 0°C) | 29.92 |
| atmospheres | kgs/sq cm | 1.0333 |
| atmospheres | kgs/sq meter | 10,332. |
| atmospheres | pounds/sq in. | 14.70 |
| atmospheres | tons/sq ft | 1.058 |
| **B** | | |
| barrels (U.S., dry) | cu. inches | 7056. |
| barrels (U.S., dry) | quarts (dry) | 105.0 |
| barrels (U.S., liquid) | gallons | 31.5 |
| barrels (oil) | gallons (oil) | 42.0 |
| bars | atmospheres | 0.9869 |
| bars | dynes/sq cm | $10^6$ |
| bars | kgs/sq meter | $1.020 \times 10^4$ |
| bars | pounds/sq ft | 2,089. |
| bars | pounds/sq in. | 14.50 |
| baryl | Dyne/sq. cm. | 1.000 |
| bolt (US Cloth) | Meters | 36.576 |
| BTU | Liter—Atmosphere | 10.409 |

* Courtesy of The Wire Association, Inc.

# CONVERSION FACTOR TABLE*

| TO CONVERT | INTO | MULTIPLY BY |
|---|---|---|
| BtU | ergs | $1.0550 \times 10^{10}$ |
| BtU | foot-lbs | 778.3 |
| BtU | gram-calories | 252.0 |
| BtU | horsepower-hrs | $3.931 \times 10^{-4}$ |
| BtU | joules | 1,054.8 |
| BtU | kilogram-calories | 0.2520 |
| BtU | kilogram-meters | 107.5 |
| BtU | kilowatt-hrs | $2.928 \times 10^{-4}$ |
| BtU/hr | foot-pounds/sec | 0.2162 |
| BtU/hr | gram-cal/sec | 0.0700 |
| BtU/hr | horsepower-hrs | $3.929 \times 10^{-4}$ |
| BtU/hr | watts | 0.2931 |
| BtU/min | foot-lbs/sec | 12.96 |
| BtU/min | horsepower | 0.02356 |
| BtU/min | kilowatts | 0.01757 |
| BtU/min | watts | 17.57 |
| BtU/sq ft/min | watts/sq in. | 0.1221 |
| bucket (Br. dry) | Cubic Cm. | $1.818 \times 10^4$ |
| bushels | cu ft | 1.2445 |
| bushels | cu in. | 2,150.4 |
| bushels | cu meters | 0.03524 |
| bushels | liters | 35.24 |
| bushels | pecks | 4.0 |
| bushels | pints (dry) | 64.0 |
| bushels | quarts (dry) | 32.0 |

## C

| TO CONVERT | INTO | MULTIPLY BY |
|---|---|---|
| calories, gram (mean) | B.T.U. (mean) | $3.9685 \times 10^{-3}$ |
| candle/sq. cm | Lamberts | 3.142 |
| candle/sq. inch | Lamberts | .4870 |
| centares (centiares) | sq meters | 1.0 |
| celsius (centigrade) | Fahrenheit | $(C° \times 9/5) + 32$ |
| centigrams | grams | 0.01 |
| centiliter | Ounce fluid (US) | .3382 |
| centiliter | Cubic inch | .6103 |
| centiliter | drams | 2.705 |
| centiliters | liters | 0.01 |
| centimeters | feet | $3.281 \times 10^{-2}$ |
| centimeters | inches | 0.3937 |
| centimeters | kilometers | $10^{-5}$ |
| centimeters | meters | 0.01 |
| centimeters | miles | $6.214 \times 10^{-6}$ |
| centimeters | millimeters | 10.0 |
| centimeters | mils | 393.7 |
| centimeters | yards | $1.094 \times 10^{-2}$ |
| centimeter-dynes | cm-grams | $1.020 \times 10^{-3}$ |
| centimeter-dynes | meter-kgs | $1.020 \times 10^{-8}$ |
| centimeter-dynes | pound-feet | $7.376 \times 10^{-8}$ |
| centimeter-grams | cm-dynes | 980.7 |
| centimeter-grams | meter-kgs | $10^{-5}$ |
| centimeter-grams | pound-feet | $7.233 \times 10^{-5}$ |
| centimeters of mercury | atmospheres | 0.01316 |
| centimeters of mercury | feet of water | 0.4461 |
| centimeters of mercury | kgs/sq meter | 136.0 |
| centimeters of mercury | pounds/sq ft | 27.85 |
| centimeters of mercury | pounds/sq in. | 0.1934 |
| centimeters/sec | feet/min | 1.1969 |
| centimeters/sec | feet/sec | 0.03281 |
| centimeters/sec | kilometers/hr | 0.036 |
| centimeters/sec | knots | 0.1943 |
| centimeters/sec | meters/min | 0.6 |
| centimeters/sec | miles/hr | 0.02237 |
| centimeters/sec | miles/min | $3.728 \times 10^{-4}$ |

* Courtesy of The Wire Association, Inc.

# CONVERSION FACTOR TABLE*

| TO CONVERT | INTO | MULTIPLY BY |
|---|---|---|
| centimeters/sec/sec | feet/sec/sec | 0.03281 |
| centimeters/sec/sec | kms/hr/sec | 0.036 |
| centimeters/sec/sec | meters/sec/sec | 0.01 |
| centimeters/sec/sec | miles/hr/sec | 0.02237 |
| chain | Inches | 792.00 |
| chain | meters | 20.12 |
| chains (surveyors' or Gunter's) | yards | 22.00 |
| circular mils | sq cms | $5.067 \times 10^{-6}$ |
| circular mils | sq mils | 0.7854 |
| Circumference | Radians | 6.283 |
| circular mils | sq incnes | $7.854 \times 10^{-7}$ |
| cords | cord feet | 8 |
| cord feet | cu. feet | 16 |
| coulomb | Statcoulombs | $2.998 \times 10^9$ |
| coulombs | faradays | $1.036 \times 10^{-5}$ |
| coulombs/sq cm | coulombs/sq in. | 64.52 |
| coulombs/sq cm | coulombs/sq meter | $10^4$ |
| coulombs/sq in. | coulombs/sq cm | 0.1550 |
| coulombs/sq in. | coulombs/sq meter | 1,550. |
| coulombs/sq meter | coulombs/sq cm | $10^{-4}$ |
| coulombs/sq meter | coulombs/sq in. | $6.452 \times 10^{-4}$ |
| cubic centimeters | cu feet | $3.531 \times 10^{-5}$ |
| cubic centimeters | cu inches | 0.06102 |
| cubic centimeters | cu meters | $10^{-6}$ |
| cubic centimeters | cu yards | $1.308 \times 10^{-6}$ |
| cubic centimeters | gallons (U. S. liq.) | $2.642 \times 10^{-4}$ |
| cubic centimeters | liters | 0.001 |
| cubic centimeters | pints (U.S. liq.) | $2.113 \times 10^{-3}$ |
| cubic centimeters | quarts (U.S. liq.) | $1.057 \times 10^{-3}$ |
| cubic feet | bushels (dry) | 0.8036 |
| cubic feet | cu cms | 28,320.0 |
| cubic feet | cu inches | 1,728.0 |
| cubic feet | cu meters | 0.02832 |
| cubic feet | cu yards | 0.03704 |
| cubic feet | gallons (U.S. liq.) | 7.48052 |
| cubic feet | liters | 28.32 |
| cubic feet | pints (U.S. liq.) | 59.84 |
| cubic feet | quarts (U.S. liq.) | 29.92 |
| cubic feet/min | cu cms/sec | 472.0 |
| cubic feet/min | gallons/sec | 0.1247 |
| cubic feet/min | liters/sec | 0.4720 |
| cubic feet/min | pounds of water/min | 62.43 |
| cubic feet/sec | million gals/day | 0.646317 |
| cubic feet/sec | gallons/min | 448.831 |
| cubic inches | cu cms | 16.39 |
| cubic inches | cu feet | $5.787 \times 10^{-4}$ |
| cubic inches | cu meters | $1.639 \times 10^{-5}$ |
| cubic inches | cu yards | $2.143 \times 10^{-5}$ |
| cubic inches | gallons | $4.329 \times 10^{-3}$ |
| cubic inches | liters | 0.01639 |
| cubic inches | mil-feet | $1.061 \times 10^5$ |
| cubic inches | pints (U.S. liq.) | 0.03463 |
| cubic inches | quarts (U.S. liq.) | 0.01732 |
| cubic meters | bushels (dry) | 28.38 |
| cubic meters | cu cms | $10^6$ |
| cubic meters | cu feet | 35.31 |
| cubic meters | cu inches | 61,023.0 |
| cubic meters | cu yards | 1.308 |
| cubic meters | gallons (U.S. liq.) | 264.2 |
| cubic meters | liters | 1,000.0 |
| cubic meters | pints (U.S. liq.) | 2,113.0 |

* Courtesy of The Wire Association, Inc.

# CONVERSION FACTOR TABLE*

| TO CONVERT | INTO | MULTIPLY BY |
|---|---|---|
| cubic meters | quarts (U.S. liq.) | 1,057. |
| cubic yards | cu cms | $7.646 \times 10^5$ |
| cubic yards | cu feet | 27.0 |
| cubic yards | cu inches | 46,656.0 |
| cubic yards | cu meters | 0.7646 |
| cubic yards | gallons (U.S. liq.) | 202.0 |
| cubic yards | liters | 764.6 |
| cubic yards | pints (U.S. liq.) | 1,615.9 |
| cubic yards | quarts (U.S. liq.) | 807.9 |
| cubic yards/min | cubic ft/sec | 0.45 |
| cubic yards/min | gallons/sec | 3.367 |
| cubic yards/min | liters/sec | 12.74 |

### D

| | | |
|---|---|---|
| dalton | Gram | $1.650 \times 10^{-24}$ |
| days | seconds | 86,400.0 |
| decigrams | grams | 0.1 |
| deciliters | liters | 0.1 |
| decimeters | meters | 0.1 |
| degrees (angle) | quadrants | 0.01111 |
| degrees (angle) | radians | 0.01745 |
| degrees (angle) | seconds | 3,600.0 |
| degrees/sec | radians/sec | 0.01745 |
| degrees/sec | revolutions/min | 0.1667 |
| degrees/sec | revolutions/sec | $2.778 \times 10^{-3}$ |
| dekagrams | grams | 10.0 |
| dekaliters | liters | 10.0 |
| dekameters | meters | 10.0 |
| drams (apothecaries' or troy) | ounces (avoidupois) | 0.1371429 |
| drams (apothecaries' or troy) | ounces (troy) | 0.125 |
| drams (U.S., fluid or apoth.) | cubic cm. | 3.6967 |
| drams | grams | 1.7718 |
| drams | grains | 27.3437 |
| drams | ounces | 0.0625 |
| dyne/cm | Erg/sq. millimeter | .01 |
| dyne/sq. cm. | Atmospheres | $9.869 \times 10^{-7}$ |
| dyne/sq. cm. | Inch of Mercury at 0°C | $2.953 \times 10^{-5}$ |
| dyne/sq. cm. | Inch of Water at 4°C | $4.015 \times 10^{-4}$ |
| dynes | grams | $1.020 \times 10^{-3}$ |
| dynes | joules/cm | $10^{-7}$ |
| dynes | joules/meter (newtons) | $10^{-5}$ |
| dynes | kilograms | $1.020 \times 10^{-6}$ |
| dynes | poundals | $7.233 \times 10^{-5}$ |
| dynes | pounds | $2.248 \times 10^{-6}$ |
| dynes/sq cm | bars | $10^{-6}$ |

### E

| | | |
|---|---|---|
| ell | Cm. | 114.30 |
| ell | Inches | 45 |
| em, Pica | Inch | .167 |
| em, Pica | Cm. | .4233 |
| erg/sec | Dyne — cm/sec | 1.000 |
| ergs | Btu | $9.480 \times 10^{-11}$ |
| ergs | dyne-centimeters | 1.0 |
| ergs | foot-pounds | $7.367 \times 10^{-8}$ |
| ergs | gram-calories | $0.2389 \times 10^{-7}$ |
| ergs | gram-cms | $1.020 \times 10^{-3}$ |
| ergs | horsepower-hrs | $3.7250 \times 10^{-14}$ |

*Courtesy of The Wire Association, Inc.

# CONVERSION FACTOR TABLE*

| TO CONVERT | INTO | MULTIPLY BY |
|---|---|---|
| ergs | joules | $10^{-7}$ |
| ergs | kg-calories | $2.389 \times 10^{-11}$ |
| ergs | kg-meters | $1.020 \times 10^{-8}$ |
| ergs | kilowatt-hrs | $0.2778 \times 10^{-13}$ |
| ergs | watt-hours | $0.2778 \times 10^{-10}$ |
| ergs/sec | Btu/min | $5,688 \times 10^{-9}$ |
| ergs/sec | ft-lbs/min | $4.427 \times 10^{-6}$ |
| ergs/sec | ft-lbs/sec | $7.3756 \times 10^{-8}$ |
| ergs/sec | horsepower | $1.341 \times 10^{-10}$ |
| ergs/sec | kg-calories/min | $1.433 \times 10^{-9}$ |
| ergs/sec | kilowatts | $10^{-10}$ |

**F**

| TO CONVERT | INTO | MULTIPLY BY |
|---|---|---|
| fahrenheit | centigrade (celsius)($F° \times 5/9$) | |
| farads | microfarads | $10^6$ |
| faraday/sec | Ampere (absolute) | $9.6500 \times 10^4$ |
| faradays | ampere-hours | 26.80 |
| faradays | coulombs | $9.649 \times 10^4$ |
| fathom | Meter | 1.828804 |
| fathoms | feet | 6.0 |
| feet | centimeters | 30.48 |
| feet | kilometers | $3.048 \times 10^{-4}$ |
| feet | meters | 0.3048 |
| feet | miles (naut.) | $1.645 \times 10^{-4}$ |
| feet | miles (stat.) | $1.894 \times 10^{-4}$ |
| feet | millimeters | 304.8 |
| feet | mils | $1.2 \times 10^4$ |
| feet of water | atmospheres | 0.02950 |
| feet of water | in. of mercury | 0.8826 |
| feet of water | kgs/sq cm | 0.03048 |
| feet of water | kgs/sq meter | 304.8 |
| feet of water | pounds/sq ft | 62.43 |
| feet of water | pounds/sq in. | 0.4335 |
| feet/min | cms/sec | 0.5080 |
| feet/min | feet/sec | 0.01667 |
| feet/min | kms/hr | 0.01829 |
| feet/min | meters/min | 0.3048 |
| feet/min | miles/hr | 0.01136 |
| feet/sec | cms/sec | 30.48 |
| feet/sec | kms/hr | 1.097 |
| feet/sec | knots | 0.5921 |
| feet/sec | meters/min | 18.29 |
| feet/sec | miles/hr | 0.6818 |
| feet/sec | miles/min | 0.01136 |
| feet/sec/sec | cms/sec/sec | 30.48 |
| feet/sec/sec | kms/hr/sec | 1.097 |
| feet/sec/sec | meters/sec/sec | 0.3048 |
| feet/sec/sec | miles/hr/sec | 0.6818 |
| feet/100 feet | per cent grade | 1.0 |
| foot – candle | Lumen/sq. meter | 10.764 |
| foot-pounds | Btu | $1.286 \times 10^{-3}$ |
| foot-pounds | ergs | $1.356 \times 10^7$ |
| foot-pounds | gram-calories | 0.3238 |
| foot-pounds | hp-hrs | $5.050 \times 10^{-7}$ |
| foot-pounds | joules | 1.356 |
| foot-pounds | kg-calories | $3.24 \times 10^{-4}$ |
| foot-pounds | kg-meters | 0.1383 |
| foot-pounds | kilowatt-hrs | $3.766 \times 10^{-7}$ |
| foot-pounds/min | Btu/min | $1.286 \times 10^{-3}$ |
| foot-pounds/min | foot-pounds/sec | 0.01667 |
| foot-pounds/min | horsepower | $3.030 \times 10^{-5}$ |
| foot-pounds/min | kg-calories/min | $3.24 \times 10^{-4}$ |

* Courtesy of The Wire Association, Inc.

# CONVERSION FACTOR TABLE*

| TO CONVERT | INTO | MULTIPLY BY |
|---|---|---|
| foot-pounds/min | kilowatts | $2.260 \times 10^{-5}$ |
| foot-pounds/sec | Btu/hr | 4.6263 |
| foot-pounds/sec | Btu/min | 0.07717 |
| foot-pounds/sec | horsepower | $1.818 \times 10^{-3}$ |
| foot-pounds/sec | kg-calories/min | 0.01945 |
| foot-pounds/sec | kilowatts | $1.356 \times 10^{-3}$ |
| furlongs | miles (U.S.) | 0.125 |
| furlongs | rods | 40.0 |
| furlongs | feet | 660.0 |

## G

| | | |
|---|---|---|
| gallons | cu cms | 3,785.0 |
| gallons | cu feet | 0.1337 |
| gallons | cu inches | 231.0 |
| gallons | cu meters | $3.785 \times 10^{-3}$ |
| gallons | cu yards | $4.951 \times 10^{-3}$ |
| gallons | liters | 3.785 |
| gallons (liq. Br. Imp.) | gallons (U.S. liq.) | 1.20095 |
| gallons (U.S.) | gallons (Imp.) | 0.83267 |
| gallons of water | pounds of water | 8.3453 |
| gallons/min | cu ft/sec | $2.228 \times 10^{-3}$ |
| gallons/min | liters/sec | 0.06308 |
| gallons/min | cu ft/hr | 8.0208 |
| gausses | lines/sq in. | 6.452 |
| gausses | webers/sq cm | $10^{-8}$ |
| gausses | webers/sq in. | $6.452 \times 10^{-8}$ |
| gausses | webers/sq meter | $10^{-4}$ |
| gilberts | ampere-turns | 0.7958 |
| gilberts/cm | amp-turns/cm | 0.7958 |
| gilberts/cm | amp-turns/in | 2.021 |
| gilberts/cm | amp-turns/meter | 79.58 |
| gills (British) | cubic cm. | 142.07 |
| gills | liters | 0.1183 |
| gills | pints (liq.) | 0.25 |
| grade | Radian | .01571 |
| grains | drams (avoirdupois) | 0.03657143 |
| grains (troy) | grains (avdp) | 1.0 |
| grains (troy) | grams | 0.06480 |
| grains (troy) | ounces (avdp) | $2.0833 \times 10^{-3}$ |
| grains (troy) | pennyweight (troy) | 0.04167 |
| grains/U.S. gal | parts/million | 17.118 |
| grains/U.S. gal | pounds/million gal | 142.86 |
| grains/Imp. gal | parts/million | 14.286 |
| grams | dynes | 980.7 |
| grams | grains | 15.43 |
| grams | joules/cm | $9.807 \times 10^{-5}$ |
| grams | joules/meter (newtons) | $9.807 \times 10^{-3}$ |
| grams | kilograms | 0.001 |
| grams | milligrams | 1,000. |
| grams | ounces (avdp) | 0.03527 |
| grams | ounces (troy) | 0.03215 |
| grams | poundals | 0.07093 |
| grams | pounds | $2.205 \times 10^{-3}$ |
| grams/cm | pounds/inch | $5.600 \times 10^{-3}$ |
| grams/cu cm | pounds/cu ft | 62.43 |
| grams/cu cm | pounds/cu in | 0.03613 |
| grams/cu cm | pounds/mil-foot | $3.405 \times 10^{-7}$ |
| grams/liter | grains/gal | 58.417 |
| grams/liter | pounds/1,000 gal | 8.345 |
| grams/liter | pounds/cu ft | 0.062427 |

* Courtesy of The Wire Association, Inc.

# CONVERSION FACTOR TABLE*

| TO CONVERT | INTO | MULTIPLY BY |
|---|---|---|
| grams/liter | parts/million | 1,000.0 |
| grams/sq cm | pounds/sq ft | 2.0481 |
| gram-calories | Btu | $3.9683 \times 10^{-3}$ |
| gram-calories | ergs | $4.1868 \times 10^7$ |
| gram-calories | foot-pounds | 3.0880 |
| gram-calories | horsepower-hrs | $1.5596 \times 10^{-6}$ |
| gram-calories | kilowatt-hrs | $1.1630 \times 10^{-6}$ |
| gram-calories | watt-hrs | $1.1630 \times 10^{-3}$ |
| gram-calories/sec | Btu/hr | 14.286 |
| gram-centimeters | Btu | $9.297 \times 10^{-8}$ |
| gram-centimeters | ergs | 980.7 |
| gram-centimeters | joules | $9.807 \times 10^{-5}$ |
| gram-centimeters | kg-cal | $2.343 \times 10^{-8}$ |
| gram-centimeters | kg-meters | $10^{-5}$ |

## H

| TO CONVERT | INTO | MULTIPLY BY |
|---|---|---|
| hand | Cm. | 10.16 |
| hectares | acres | 2.471 |
| henries | sq feet | $1.076 \times 10^5$ |
| hectograms | grams | 100.0 |
| hectoliters | liters | 100.0 |
| hectometers | meters | 100.0 |
| hectowatts | watts | 100.0 |
| henries | millihenries | 1,000.0 |
| hogsheads(British) | cubic ft. | 10.114 |
| hogsheads (U.S.) | cubic ft. | 8.42184 |
| hogsheads (U.S.) | gallons (U.S.) | 63 |
| horsepower | Btu/min | 42.44 |
| horsepower | foot-lbs/min | 33,000. |
| horsepower | foot-lbs/sec | 550.0 |
| horsepower (metric) (542.5 ft lb/sec) | horsepower (550 ft lb/sec) | 0.9863 |
| horsepower (550 ft lb/sec) | horsepower (metric) (542.5 ft lb/sec) | 1.014 |
| horsepower | kg-calories/min | 10.68 |
| horsepower | kilowatts | 0.7457 |
| horsepower | watts | 745.7 |
| horsepower (boiler) | Btu/hr | 33,479 |
| horsepower (boiler) | kilowatts | 9.803 |
| horsepower-hrs | Btu | 2,547. |
| horsepower-hrs | ergs | $2.6845 \times 10^{13}$ |
| horsepower-hrs | foot-lbs | $1.98 \times 10^6$ |
| horsepower-hrs | gram-calories | 641,190. |
| horsepower-hrs | joules | $2.684 \times 10^6$ |
| horsepower-hrs | kg-calories | 641.1 |
| horsepower-hrs | kg-meters | $2.737 \times 10^5$ |
| horsepower-hrs | kilowatt-hrs | 0.7457 |
| hours | days | $4.167 \times 10^{-2}$ |
| hours | weeks | $5.952 \times 10^{-3}$ |
| hundredweights (long) | pounds | 112 |
| hundredweights (long) | tons (long) | 0.05 |
| hundredweights (short) | ounces (avoirdupois) | 1600 |
| hundredweights (short) | pounds | 100 |
| hundredweights (short) | tons (metric) | 0.0453592 |
| hundredweights (short) | tons (long) | 0.0446429 |

## I

| TO CONVERT | INTO | MULTIPLY BY |
|---|---|---|
| inches | centimeters | 2.540 |
| inches | meters | $2.540 \times 10^{-2}$ |
| inches | miles | $1.578 \times 10^{-5}$ |
| inches | millimeters | 25.40 |

* Courtesy of The Wire Association, Inc.

# CONVERSION FACTOR TABLE*

| TO CONVERT | INTO | MULTIPLY BY |
|---|---|---|
| inches | mils | 1,000.0 |
| inches | yards | $2.778 \times 10^{-2}$ |
| inches of mercury | atmospheres | 0.03342 |
| inches of mercury | feet of water | 1.133 |
| inches of mercury | kgs/sq cm | 0.03453 |
| inches of mercury | kgs/sq meter | 345.3 |
| inches of mercury | pounds/sq ft | 70.73 |
| inches of mercury | pounds/sq in. | 0.4912 |
| inches of water (at 4°C) | atmospheres | $2.458 \times 10^{-3}$ |
| inches of water (at 4°C) | inches of mercury | 0.07355 |
| inches of water (at 4°C) | kgs/sq cm | $2.540 \times 10^{-3}$ |
| inches of water (at 4°C) | ounces/sq in. | 0.5781 |
| inches of water (at 4°C) | pounds/sq ft | 5.204 |
| inches of water (at 4°C) | pounds/sq in. | 0.03613 |
| International Ampere | Ampere (absolute) | .9998 |
| International Volt | Volts (absolute) | 1.0003 |
| International volt | Joules (absolute) | $1-593 \times 10^{-19}$ |
| International volt | Joules | $9.654 \times 10^{4}$ |

## J

| | | |
|---|---|---|
| joules | Btu | $9.480 \times 10^{-4}$ |
| joules | ergs | $10^{7}$ |
| joules | foot-pounds | 0.7376 |
| joules | kg-calories | $2.389 \times 10^{-4}$ |
| joules | kg-meters | 0.1020 |
| joules | watt-hrs | $2.778 \times 10^{-4}$ |
| joules/cm | grams | $1.020 \times 10^{4}$ |
| joules/cm | dynes | $10^{7}$ |
| joules/cm | joules/meter (newtons) | 100.0 |
| joules/cm | poundals | 723.3 |
| joules/cm | pounds | 22.48 |

## K

| | | |
|---|---|---|
| kilograms | dynes | 980,665. |
| kilograms | grams | 1,000.0 |
| kilograms | joules/cm | 0.09807 |
| kilograms | joules/meter (newtons) | 9.807 |
| kilograms | poundals | 70.93 |
| kilograms | pounds | 2.205 |
| kilograms | tons (long) | $9.842 \times 10^{-4}$ |
| kilograms | tons (short) | $1.102 \times 10^{-3}$ |
| kilograms/cu meter | grams/cu cm | 0.001 |
| kilograms/cu meter | pounds/cu ft | 0.06243 |
| kilograms/cu meter | pounds/cu in. | $3.613 \times 10^{-5}$ |
| kilograms/cu meter | pounds/mil-foot | $3.405 \times 10^{-10}$ |
| kilograms/meter | pounds/ft | 0.6720 |
| Kilogram/sq. cm. | Dynes | 980,665 |
| kilograms/sq cm | atmospheres | 0.9678 |
| kilograms/sq cm | feet of water | 32.81 |
| kilograms/sq cm | inches of mercury | 28.96 |
| kilograms/sq cm | pounds/sq ft | 2,048. |
| kilograms/sq cm | pounds/sq in. | 14.22 |
| kilograms/sq meter | atmospheres | $9.678 \times 10^{-5}$ |
| kilograms/sq meter | bars | $98.07 \times 10^{-6}$ |
| kilograms/sq meter | feet of water | $3.281 \times 10^{-3}$ |
| kilograms/sq meter | inches of mercury | $2.896 \times 10^{-3}$ |
| kilograms/sq meter | pounds/sq ft | 0.2048 |
| kilograms/sq meter | pounds/sq in. | $1.422 \times 10^{-3}$ |
| kilograms/sq mm | kgs/sq meter | $10^{6}$ |

* Courtesy of The Wire Association, Inc.

# CONVERSION FACTOR TABLE*

| TO CONVERT | INTO | MULTIPLY BY |
|---|---|---|
| kilogram-calories | Btu | 3.968 |
| kilogram-calories | foot-pounds | 3,088. |
| kilogram-calories | hp-hrs | $1.560 \times 10^{-3}$ |
| kilogram-calories | joules | 4,186. |
| kilogram-calories | kg-meters | 426.9 |
| kilogram-calories | kilojoules | 4.186 |
| kilogram-calories | kilowatt-hrs | $1.163 \times 10^{-3}$ |
| kilogram meters | Btu | $9.294 \times 10^{-3}$ |
| kilogram meters | ergs | $9.804 \times 10^{7}$ |
| kilogram meters | foot-pounds | 7.233 |
| kilogram meters | joules | 9.804 |
| kilogram meters | kg-calories | $2.342 \times 10^{-3}$ |
| kilogram meters | kilowatt-hrs | $2.723 \times 10^{-6}$ |
| kilolines | maxwells | 1,000.0 |
| kiloliters | liters | 1,000.0 |
| kilometers | centimeters | $10^5$ |
| kilometers | feet | 3,281. |
| kilometers | inches | $3.937 \times 10^4$ |
| kilometers | meters | 1,000.0 |
| kilometers | miles | 0.6214 |
| kilometers | millimeters | $10^6$ |
| kilometers | yards | 1,094. |
| kilometers/hr | cms/sec | 27.78 |
| kilometers/hr | feet/min | 54.68 |
| kilometers/hr | feet/sec | 0.9113 |
| kilometers/hr | knots | 0.5396 |
| kilometers/hr | meters/min | 16.67 |
| kilometers/hr | miles/hr | 0.6214 |
| kilometers/hr/sec | cms/sec/sec | 27.78 |
| kilometers/hr/sec | ft/sec/sec | 0.9113 |
| kilometers/hr/sec | meters/sec/sec | 0.2778 |
| kilometers/hr/sec | miles/hr/sec | 0.6214 |
| kilowatts | Btu/min | 56.92 |
| kilowatts | foot-lbs/min | $4.426 \times 10^4$ |
| kilowatts | foot-lbs/sec | 737.6 |
| kilowatts | horsepower | 1.341 |
| kilowatts | kg-calories/min | 14.34 |
| kilowatts | watts | 1,000.0 |
| kilowatt-hrs | Btu | 3,413. |
| kilowatt-hrs | ergs | $3.600 \times 10^{13}$ |
| kilowatt-hrs | foot-lbs | $2.655 \times 10^6$ |
| kilowatt-hrs | gram-calories | 859,850. |
| kilowatt-hrs | horsepower-hrs | 1.341 |
| kilowatt-hrs | joules | $3.6 \times 10^6$ |
| kilowatt-hrs | kg-calories | 860.5 |
| kilowatt-hrs | kg-meters | $3.671 \times 10^5$ |
| kilowatt-hrs | pounds of water evaporated from and at 212° F. | 3.53 |
| kilowatt-hrs | pounds of water raised from 62° to 212° F. | 22.75 |
| knots | feet/hr | 6,080. |
| knots | kilometers/hr | 1.8532 |
| knots | nautical miles/hr | 1.0 |
| knots | statute miles/hr | 1.151 |
| knots | yards/hr | 2,027. |
| knots | feet/sec | 1.689 |

* Courtesy of The Wire Association, Inc.

# CONVERSION FACTOR TABLE*

| TO CONVERT | INTO | MULTIPLY BY |
|---|---|---|

## L

| | | |
|---|---|---|
| league | miles (approx.) | 3.0 |
| light year | Miles | $5.9 \times 10^{12}$ |
| light year | Kilometers | $9.46091 \times 10^{12}$ |
| lines/sq cm | gausses | 1.0 |
| lines/sq in. | gausses | 0.1550 |
| lines/sq in. | webers/sq cm | $1.550 \times 10^{-9}$ |
| lines/sq in. | webers/sq in. | $10^{-8}$ |
| lines/sq in. | webers/sq meter | $1.550 \times 10^{-5}$ |
| links (engineer's) | inches | 12.0 |
| links (surveyor's) | inches | 7.92 |
| liters | bushels (U.S. dry) | 0.02838 |
| liters | cu cm | 1,000.0 |
| liters | cu feet | 0.03531 |
| liters | cu inches | 61.02 |
| liters | cu meters | 0.001 |
| liters | cu yards | $1.308 \times 10^{-3}$ |
| liters | gallons (U.S. liq.) | 0.2642 |
| liters | pints (U.S. liq.) | 2.113 |
| liters | quarts (U.S. liq.) | 1.057 |
| liters/min | cu ft/sec | $5.886 \times 10^{-4}$ |
| liters/min | gals/sec | $4.403 \times 10^{-3}$ |
| lumens/sq ft | foot-candles | 1.0 |
| lumen | Spherical candle power | .07958 |
| lumen | Watt | .001496 |
| lumen/sq. ft. | Lumen/sq. meter | 10.76 |
| lux | foot-candles | 0.0929 |

## M

| | | |
|---|---|---|
| maxwells | kilolines | 0.001 |
| maxwells | webers | $10^{-8}$ |
| megalines | maxwells | $10^{6}$ |
| megohms | microhms | $10^{12}$ |
| megohms | ohms | $10^{6}$ |
| meters | centimeters | 100.0 |
| meters | feet | 3.281 |
| meters | inches | 39.37 |
| meters | kilometers | 0.001 |
| meters | miles (naut.) | $5.396 \times 10^{-4}$ |
| meters | miles (stat.) | $6.214 \times 10^{-4}$ |
| meters | millimeters | 1,000.0 |
| meters | yards | 1.094 |
| meters | varas | 1.179 |
| meters/min | cms/sec | 1.667 |
| meters/min | feet/min | 3.281 |
| meters/min | feet/sec | 0.05468 |
| meters/min | kms/hr | 0.06 |
| meters/min | knots | 0.03238 |
| meters/min | miles/hr | 0.03728 |
| meters/sec | feet/min | 196.8 |
| meters/sec | feet/sec | 3.281 |
| meters/sec | kilometers/hr | 3.6 |
| meters/sec | kilometers/min | 0.06 |
| meters/sec | miles/hr | 2.237 |
| meters/sec | miles/min | 0.03728 |
| meters/sec/sec | cms/sec/sec | 100.0 |
| meters/sec/sec | ft/sec/sec | 3.281 |
| meters/sec/sec | kms/hr/sec | 3.6 |
| meters/sec/sec | miles/hr/sec | 2.237 |
| meter-kilograms | cm-dynes | $9.807 \times 10^{7}$ |
| meter-kilograms | cm-grams | $10^{5}$ |

* Courtesy of The Wire Association, Inc.

# CONVERSION FACTOR TABLE*

| TO CONVERT | INTO | MULTIPLY BY |
|---|---|---|
| meter-kilograms | pound-feet | 7.233 |
| microfarad | farads | $10^{-6}$ |
| micrograms | grams | $10^{-6}$ |
| microhms | megohms | $10^{-12}$ |
| microhms | ohms | $10^{-6}$ |
| microliters | liters | $10^{-6}$ |
| microns | meters | $1 \times 10^{-6}$ |
| miles (naut.) | feet | 6,080.27 |
| miles (naut.) | kilometers | 1.853 |
| miles (naut.) | meters | 1,853. |
| miles (naut.) | miles (statute) | 1.1516 |
| miles (naut.) | yards | 2,027. |
| miles (statute) | centimeters | $1.609 \times 10^5$ |
| miles (statute) | feet | 5,280. |
| miles (statute) | inches | $6.336 \times 10^4$ |
| miles (statute) | kilometers | 1.609 |
| miles (statute) | meters | 1,609. |
| miles (statute) | miles (naut.) | 0.8684 |
| miles (statute) | yards | 1,760. |
| miles/hr | cms/sec | 44.70 |
| miles/hr | feet/min | 88. |
| miles/hr | feet/sec | 1.467 |
| miles/hr | kms/hr | 1.609 |
| miles/hr | kms/min | 0.02682 |
| miles/hr | knots | 0.8684 |
| miles/hr | meters/min | 26.82 |
| miles/hr | miles/min | 0.1667 |
| miles/hr/sec | cms/sec/sec | 44.70 |
| miles/hr/sec | feet/sec/sec | 1.467 |
| miles/hr/sec | kms/hr/sec | 1.609 |
| miles/hr/sec | meters/sec/sec | 0.4470 |
| miles/min | cms/sec | 2,682. |
| miles/min | feet/sec | 88. |
| miles/min | kms/min | 1.609 |
| miles/min | knots/min | 0.8684 |
| miles/min | miles/hr | 60.0 |
| mil-feet | cu inches | $9.425 \times 10^{-6}$ |
| milliers | kilograms | 1,000. |
| millimicrons | meters | $1 \times 10^{-9}$ |
| milligrams | grains | 0.01543236 |
| milligrams | grams | 0.001 |
| milligrams/liter | parts/million | 1.0 |
| millihenries | henries | 0.001 |
| milliliters | liters | 0.001 |
| millimeters | centimeters | 0.1 |
| millimeters | feet | $3.281 \times 10^{-3}$ |
| millimeters | inches | 0.03937 |
| millimeters | kilometers | $10^{-6}$ |
| millimeters | meters | 0.001 |
| millimeters | miles | $6.214 \times 10^{-7}$ |
| millimeters | mils | 39.37 |
| millimeters | yards | $1.094 \times 10^{-3}$ |
| million gals/day | cu ft/sec | 1.54723 |
| mils | centimeters | $2.540 \times 10^{-3}$ |
| mils | feet | $8.333 \times 10^{-5}$ |
| mils | inches | 0.001 |
| mils | kilometers | $2.540 \times 10^{-8}$ |
| mils | yards | $2.778 \times 10^{-5}$ |
| miner's inches | cu ft/min | 1.5 |
| minims (British) | cubic cm. | 0.059192 |
| minims (U.S., fluid) | cubic cm. | 0.061612 |
| minutes (angles) | degrees | 0.01667 |

* Courtesy of The Wire Association, Inc.

# CONVERSION FACTOR TABLE*

| TO CONVERT | INTO | MULTIPLY BY |
|---|---|---|
| minutes (angles) | quadrants | $1.852 \times 10^{-4}$ |
| minutes (angles) | radians | $2.909 \times 10^{-4}$ |
| minutes (angles) | seconds | 60.0 |
| myriagrams | kilograms | 10.0 |
| myriameters | kilometers | 10.0 |
| myriawatts | kilowatts | 10.0 |

### N

| | | |
|---|---|---|
| nepers | decibels | 8.686 |
| newton | Dynes | $1 \times 105$ |

### O

| | | |
|---|---|---|
| ohm (International) | ohm (absolute) | 1.0005 |
| ohms | megohms | $10^{-6}$ |
| ohms | microhms | $10^6$ |
| ounces | drams | 16.0 |
| ounces | grains | 437.5 |
| ounces | grams | 28.349527 |
| ounces | pounds | 0.0625 |
| ounces | ounces (troy) | 0.9115 |
| ounces | tons (long) | $2.790 \times 10^{-5}$ |
| ounces | tons (metric) | $2.835 \times 10^{-5}$ |
| ounces (fluid) | cu inches | 1.805 |
| ounces (fluid) | liters | 0.02957 |
| ounces (troy) | grains | 480.0 |
| ounces (troy) | grams | 31.103481 |
| ounces (troy) | ounces (avdp.) | 1.09714 |
| ounces (troy) | pennyweights (troy) | 20.0 |
| ounces (troy) | pounds (troy) | 0.08333 |
| ounce/sq. inch | Dynes/sq. cm. | 4309 |
| ounces/sq in. | pounds/sq in. | 0.0625 |

### P

| | | |
|---|---|---|
| parsec | Miles | $19 \times 10^{12}$ |
| parsec | Kilometers | $3.084 \times 10^{13}$ |
| parts/million | grains/U.S. gal | 0.0584 |
| parts/million | grains/Imp. gal | 0.07016 |
| parts/million | pounds/million gal | 8.345 |
| pecks (British) | cubic inches | 554.6 |
| pecks (British) | liters | 9.091901 |
| pecks (U.S.) | bushels | 0.25 |
| pecks (U.S.) | cubic inches | 537.605 |
| pecks (U.S.) | liters | 8.809582 |
| pecks (U.S.) | quarts (dry) | 8 |
| pennyweights (troy) | grains | 24.0 |
| pennyweights (troy) | ounces (troy) | 0.05 |
| pennyweights (troy) | grams | 1.55517 |
| pennyweights (troy) | pounds (troy) | $4.1667 \times 10^{-3}$ |
| pints (dry) | cu inches | 33.60 |
| pints (liq.) | cu cms. | 473.2 |
| pints (liq.) | cu feet | 0.01671 |
| pints (liq.) | cu inches | 28.87 |
| pints (liq.) | cu meters | $4.732 \times 10^{-4}$ |
| pints (liq.) | cu yards | $6.189 \times 10^{-4}$ |
| pints (liq.) | gallons | 0.125 |
| pints (liq.) | liters | 0.4732 |
| pints (liq.) | quarts (liq.) | 0.5 |
| planck's quantum | Erg − second | $6.624 \times 10^{-27}$ |
| poise | Gram/cm. sec. | 1.00 |
| pounds (avoirdupois) | ounces (troy) | 14.5833 |

* Courtesy of The Wire Association, Inc.

# CONVERSION FACTOR TABLE*

| TO CONVERT | INTO | MULTIPLY BY |
|---|---|---|
| poundals | dynes | 13,826. |
| poundals | grams | 14.10 |
| poundals | joules/cm | $1.383 \times 10^{-3}$ |
| poundals | joules/meter (newtons) | 0.1383 |
| poundals | kilograms | 0.01410 |
| poundals | pounds | 0.03108 |
| pounds | drams | 256. |
| pounds | dynes | $44.4823 \times 10^4$ |
| pounds | grains | 7,000. |
| pounds | grams | 453.5924 |
| pounds | joules/cm | 0.04448 |
| pounds | joules/meter (newtons) | 4.448 |
| pounds | kilograms | 0.4536 |
| pounds | ounces | 16.0 |
| pounds | ounces (troy) | 14.5833 |
| pounds | poundals | 32.17 |
| pounds | pounds (troy) | 1.21528 |
| pounds | tons (short) | 0.0005 |
| pounds (troy) | grains | 5,760. |
| pounds (troy) | grams | 373.24177 |
| pounds (troy) | ounces (avdp.) | 13.1657 |
| pounds (troy) | ounces (troy) | 12.0 |
| pounds (troy) | pennyweights (troy) | 240.0 |
| pounds (troy) | pounds (avdp.) | 0.822857 |
| pounds (troy) | tons (long) | $3.6735 \times 10^{-4}$ |
| pounds (troy) | tons (metric) | $3.7324 \times 10^{-4}$ |
| pounds (troy) | tons (short) | $4.1143 \times 10^{-4}$ |
| pounds of water | cu feet | 0.01602 |
| pounds of water | cu inches | 27.68 |
| pounds of water | gallons | 0.1198 |
| pounds of water/min | cu ft/sec | $2.670 \times 10^{-4}$ |
| pound-feet | cm-dynes | $1.356 \times 10^7$ |
| pound-feet | cm-grams | 13,825. |
| pound-feet | meter-kgs | 0.1383 |
| pounds/cu ft | grams/cu cm | 0.01602 |
| pounds/cu ft | kgs/cu meter | 16.02 |
| pounds/cu ft | pounds/cu in. | $5.787 \times 10^{-4}$ |
| pounds/cu ft | pounds/mil-foot | $5.456 \times 10^{-9}$ |
| pounds/cu in. | gms/cu cm | 27.68 |
| pounds/cu in. | kgs/cu meter | $2.768 \times 10^4$ |
| pounds/cu in. | pounds/cu ft | 1,728. |
| pounds/cu in. | pounds/mil-foot | $9.425 \times 10^{-6}$ |
| pounds/ft | kgs/meter | 1.488 |
| pounds/in. | gms/cm | 178.6 |
| pounds/mil-foot | gms/cu cm | $2.306 \times 10^6$ |
| pounds/sq ft | atmospheres | $4.725 \times 10^{-4}$ |
| pounds/sq ft | feet of water | 0.01602 |
| pounds/sq ft | inches of mercury | 0.01414 |
| pounds/sq ft | kgs/sq meter | 4.882 |
| pounds/sq ft | pounds/sq in. | $6.944 \times 10^{-3}$ |
| pounds/sq in. | atmospheres | 0.06804 |
| pounds/sq in. | feet of water | 2.307 |
| pounds/sq in. | inches of mercury | 2.036 |
| pounds/sq in. | kgs/sq meter | 703.1 |
| pounds/sq in. | pounds/sq ft | 144.0 |

## Q

| | | |
|---|---|---|
| quadrants (angle) | degrees | 90.0 |
| quadrants (angle) | minutes | 5,400.0 |
| quadrants (angle) | radians | 1.571 |
| quadrants (angle) | seconds | $3.24 \times 10^5$ |

*Courtesy of The Wire Association, Inc.

# CONVERSION FACTOR TABLE*

| TO CONVERT | INTO | MULTIPLY BY |
|---|---|---|
| quarts (dry) | cu inches | 67.20 |
| quarts (liq.) | cu cms | 946.4 |
| quarts (liq.) | cu feet | 0.03342 |
| quarts (liq.) | cu inches | 57.75 |
| quarts (liq.) | cu meters | $9.464 \times 10^{-4}$ |
| quarts (liq.) | cu yards | $1.238 \times 10^{-3}$ |
| quarts (liq.) | gallons | 0.25 |
| quarts (liq.) | liters | 0.9463 |
| **R** | | |
| radians | degrees | 57.30 |
| radians | minutes | 3,438. |
| radians | quadrants | 0.6366 |
| radians | seconds | $2.063 \times 10^{5}$ |
| radians/sec | degrees/sec | 57.30 |
| radians/sec | revolutions/min | 9.549 |
| radians/sec | revolutions/sec | 0.1592 |
| radians/sec/sec | revs/min/min | 573.0 |
| radians/sec/sec | revs/min/sec | 9.549 |
| radians/sec/sec | revs/sec/sec | 0.1592 |
| revolutions | degrees | 360.0 |
| revolutions | quadrants | 4.0 |
| revolutions | radians | 6.283 |
| revolutions/min | degrees/sec | 6.0 |
| revolutions/min | radians/sec | 0.1047 |
| revolutions/min | revs/sec | 0.01667 |
| revolutions/min/min | radians/sec/sec | $1.745 \times 10^{-3}$ |
| revolutions/min/min | revs/min/sec | 0.01667 |
| revolutions/min/min | revs/sec/sec | $2.778 \times 10^{-4}$ |
| revolutions/sec | degrees/sec | 360.0 |
| revolutions/sec | radians/sec | 6.283 |
| revolutions/sec | revs/min | 60.0 |
| revolutions/sec/sec | radians/sec/sec | 6.283 |
| revolutions/sec/sec | revs/min/min | 3,600.0 |
| revolutions/sec/sec | revs/min/sec | 60.0 |
| rod | Chain (Gunters) | .25 |
| rod | Meters | 5.029 |
| rod (Surveyors' meas.) | yards | 5.5 |
| rods | feet | 16.5 |
| **S** | | |
| scruples | grains | 20 |
| seconds (angle) | degrees | $2.778 \times 10^{-4}$ |
| seconds (angle) | minutes | 0.01667 |
| seconds (angle) | quadrants | $3.087 \times 10^{-6}$ |
| seconds (angle) | radians | $4.848 \times 10^{-6}$ |
| slug | Kilogram | 14.59 |
| slug | Pounds | 32.17 |
| sphere | Steradians | 12.57 |
| square centimeters | circular mils | $1.973 \times 10^{5}$ |
| square centimeters | sq feet | $1.076 \times 10^{-3}$ |
| square centimeters | sq inches | 0.1550 |
| square centimeters | sq meters | 0.0001 |
| square centimeters | sq miles | $3.861 \times 10^{-11}$ |
| square centimeters | sq millimeters | 100.0 |
| square centimeters | sq yards | $1.196 \times 10^{-4}$ |
| square feet | acres | $2.296 \times 10^{-5}$ |
| square feet | circular mils | $1.833 \times 10^{8}$ |
| square feet | sq cms | 929.0 |
| square feet | sq inches | 144.0 |

* Courtesy of The Wire Association, Inc.

# CONVERSION FACTOR TABLE*

| TO CONVERT | INTO | MULTIPLY BY |
|---|---|---|
| square feet | sq meters | 0.09290 |
| square feet | sq miles | $3.587 \times 10^{-8}$ |
| square feet | sq millimeters | $9.290 \times 10^4$ |
| square feet | sq yards | 0.1111 |
| square inches | circular mils | $1.273 \times 10^6$ |
| square inches | sq cms | 6.452 |
| square inches | sq feet | $6.944 \times 10^{-3}$ |
| square inches | sq millimeters | 645.2 |
| square inches | sq mils | $10^6$ |
| square inches | sq yards | $7.716 \times 10^{-4}$ |
| square kilometers | acres | 247.1 |
| square kilometers | sq cms | $10^{10}$ |
| square kilometers | sq ft | $10.76 \times 10^6$ |
| square kilometers | sq inches | $1.550 \times 10^9$ |
| square kilometers | sq meters | $10^6$ |
| square kilometers | sq miles | 0.3861 |
| square kilometers | sq yards | $1.196 \times 10^6$ |
| square meters | acres | $2.471 \times 10^{-4}$ |
| square meters | sq cms | $10^4$ |
| square meters | sq feet | 10.76 |
| square meters | sq inches | 1,550. |
| square meters | sq miles | $3.861 \times 10^{-7}$ |
| square meters | sq millimeters | $10^6$ |
| square meters | sq yards | 1.196 |
| square miles | acres | 640.0 |
| square miles | sq feet | $27.88 \times 10^6$ |
| square miles | sq kms | 2.590 |
| square miles | sq meters | $2.590 \times 10^6$ |
| square miles | sq yards | $3.098 \times 10^6$ |
| square millimeters | circular mils | 1,973. |
| square millimeters | sq cms | 0.01 |
| square millimeters | sq feet | $1.076 \times 10^{-5}$ |
| square millimeters | sq inches | $1.550 \times 10^{-3}$ |
| square mils | circular mils | 1.273 |
| square mils | sq cms | $6.452 \times 10^{-6}$ |
| square mils | sq inches | $10^{-6}$ |
| square yards | acres | $2.066 \times 10^{-4}$ |
| square yards | sq cms | 8,361. |
| square yards | sq feet | 9.0 |
| square yards | sq inches | 1,296. |
| square yards | sq meters | 0.8361 |
| square yards | sq miles | $3.228 \times 10^{-7}$ |
| square yards | sq millimeters | $8.361 \times 10^5$ |

## T

| | | |
|---|---|---|
| temperature (°C) +273 | absolute temperature (°C) | 1.0 |
| temperature (°C) +17.78 | temperature (°F) | 1.8 |
| temperature (°F) +460 | absolute temperature (°F) | 1.0 |
| temperature (°F) −32 | temperature (°C) | 5/9 |
| tons (long) | kilograms | 1,016. |
| tons (long) | pounds | 2,240. |
| tons (long) | tons (short) | 1.120 |
| tons (metric) | kilograms | 1,000. |
| tons (metric) | pounds | 2,205. |
| tons (short) | kilograms | 907.1848 |
| tons (short) | ounces | 32,000. |
| tons (short) | ounces (troy) | 29,166.66 |
| tons (short) | pounds | 2,000. |

* Courtesy of The Wire Association, Inc.

# CONVERSION FACTOR TABLE*

| TO CONVERT | INTO | MULTIPLY BY |
|---|---|---|
| tons (short) | pounds (troy) | 2,430.56 |
| tons (short) | tons (long) | 0.89287 |
| tons (short) | tons (metric) | 0.9078 |
| tons (short)/sq ft | kgs/sq meter | 9,765. |
| tons (short)/sq ft | pounds/sq in. | 2,000. |
| tons of water/24 hrs | pounds of water/hr | 83.333 |
| tons of water/24 hrs | gallons/min | 0.16643 |
| tons of water/24 hrs | cu ft/hr | 1.3349 |

### V

| | | |
|---|---|---|
| volt/inch | Volt/cm. | .39370 |
| volt (absolute) | Statvolts | .003336 |

### W

| | | |
|---|---|---|
| watts | Btu/hr | 3.4129 |
| watts | Btu/min | 0.05688 |
| watts | ergs/sec | 107. |
| watts | foot-lbs/min | 44.27 |
| watts | foot-lbs/sec | 0.7378 |
| watts | horsepower | $1.341 \times 10^{-3}$ |
| watts | horsepower (metric) | $1.360 \times 10^{-3}$ |
| watts | kg-calories/min | 0.01433 |
| watts | kilowatts | 0.001 |
| watts (Abs.) | B.T.U. (mean)/min. | 0.056884 |
| watts (Abs.) | joules/sec. | 1 |
| watt-hours | Btu | 3.413 |
| watt-hours | ergs | $3.60 \times 10^{10}$ |
| watt-hours | foot-pounds | 2,656. |
| watt-hours | gram-calories | 859.85 |
| watt-hours | horsepower-hrs | $1.341 \times 10^{-3}$ |
| watt-hours | kilogram-calories | 0.8605 |
| watt-hours | kilogram-meters | 367.2 |
| watt-hours | kilowatt-hrs | 0.001 |
| watt (International) | Watt (absolute) | 1.0002 |
| webers | maxwells | $10^8$ |
| webers | kilolines | $10^5$ |
| webers/sq in. | gausses | $1.550 \times 10^7$ |
| webers/sq in. | lines/sq in. | $10^8$ |
| webers/sq in. | webers/sq cm | 0.1550 |
| webers/sq in. | webers/sq meter | 1,550. |
| webers/sq meter | gausses | $10^4$ |
| webers/sq meter | lines/sq in. | $6.452 \times 10^4$ |
| webers/sq meter | webers/sq cm | $10^{-4}$ |
| webers/sq meter | webers/sq in. | $6.452 \times 10^{-4}$ |

### Y

| | | |
|---|---|---|
| yards | centimeters | 91.44 |
| yards | kilometers | $9.144 \times 10^{-4}$ |
| yards | meters | 0.9144 |
| yards | miles (naut.) | $4.934 \times 10^{-4}$ |
| yards | miles (stat.) | $5.682 \times 10^{-4}$ |
| yards | millimeters | 914.4 |

* Courtesy of The Wire Association, Inc.

# INDEX
## —A—

# INDEX
## —B—

197

# INDEX
## —C—

## INDEX
### —F—

## INDEX
### —G—

## INDEX
## —H—

## INDEX
## —I—

## INDEX
## —N—

## INDEX
## —O—

## INDEX
## —P—

## INDEX
## —R—

## INDEX
## —S—

# INDEX
# —T—

## INDEX
## — V —

## INDEX
## — W —

## INDEX
## — Z —